The
NATURE of
LIFE and DEATH

The
NATURE *of*
LIFE *and* DEATH

EVERY BODY LEAVES A TRACE

PATRICIA WILTSHIRE

G. P. Putnam's Sons
New York

PUTNAM
— EST. 1838 —

G. P. PUTNAM'S SONS

Publishers Since 1838

An imprint of Penguin Random House LLC

penguinrandomhouse.com

First published by Bonnier Books U.K.

ISBN 9780525542216

Printed in the United States of America

1   3   5   7   9   10   8   6   4   2

Book design by Elke Sigal

Penguin is committed to publishing works of quality and integrity. In that spirit, we are proud to offer this book to our readers; however, the story, experiences, and the words are the author's alone.

I dedicate this book to my dearest grandmother,
Vera May Tiley (neé Gow), who gave me much love,
and taught me how to be brave in the face of adversity.

CONTENTS

CONTENTS

The
NATURE *of*
LIFE *and* DEATH

CHAPTER 1

Beginnings

Imagine, for a moment, that you are walking in a winter wood. The ground is soft beneath your feet; something catches your eye—something untoward, something not quite natural, in a depression just off the beaten track. Perhaps you are here walking your dog (this is the way so many stories start). Perhaps the dog hurtles off into the undergrowth and whines. As you fight through the brambles to reach it, you have a moment of foreboding—and, looking down, you realize why . . . for there in front of you where the dog has frantically scraped the soil, the lifeless hand of a body is exposed, its pallor stark against the black humus.

It was not so very long ago that identifying the culprit in a crime like this might have been possible only by the testimony of witnesses or the confessions of the accused. Within living memory, and in the absence of any clues to its identity or to connect it to a potential suspect, a body discovered in a shallow grave might have remained a mystery forever. But times move on. The world of forensic detection gathers pace.

We are all familiar with the idea of fingerprints, and they have even been found in prehistoric pottery. The ancient Chinese and Assyrians used fingerprints to establish ownership of clay artifacts and, later on, documents. Sir William Herschel insisted on having fingerprints as well as signatures on civil contracts when he was a British administrator in India in 1858. Fingerprint analysis was firmly established by the late nineteenth century and, in 1882, Alphonse Bertillon, a French anthropologist, was routinely recording fingerprints on cards in his academic research into variation in people and, by 1891, the Argentine police had started fingerprinting criminals. The field developed apace and, in 1911, fingerprinting became accepted by the US courts as a reliable method of identifying individuals. Fast forward to 1980 to the first computerized database of fingerprints, NAFIS (National Automated Fingerprint Identification System), established in the United Kingdom and the United States.

In the 1990s, another quantum leap was taken in forensic detection with the development of DNA profiling. This, like fingerprinting before it, enabled the unique imprint of an individual to be captured but by taking samples of blood, semen, body cells, or roots of hairs. This development transformed the world of forensic detection, making identifying unknown victims, like our body in the winter wood, or connecting a person to a crime scene, much easier. Make no mistake about it, these were seismic moments in the history of forensic detection. Murderers who might otherwise have gone free have been imprisoned because of these advances. Rapists who would have gone on to rape again were caught and put behind bars. So too were innocent parties exonerated for the crimes of which they had been unjustly accused. Step by step, and with plenty of backslides along the way, police work got closer to the truth.

Fingerprints are not always found at a crime scene, especially if that crime is being perpetrated by somebody who is forensically aware and wears gloves, or covers their tracks behind them. And neither is DNA evidence as omnipotent or omnipresent as many people think; there might be no trace at all of an offender left behind at a crime scene—no hair, blood or semen, nor any other body fluids or tissues—so that building up a genetic profile of an assailant is just not possible.

Yet . . . what if there was another way to connect people and places, to exonerate the innocent and indicate somebody's guilt? What if, aside from fingerprint and DNA evidence, there were other traces left on us that could corroborate one version of events over another? And what if it was so pervasive that, no matter how forensically aware a criminal was, they could never quite shake it off?

Imagine, again, that you are back in that winter wood. As you push through the brambles and overhanging branches to reach the body, the sleeve of your coat rubs up against an oak tree, collecting the microscopic spores and pollen that have become impacted in the crevices of the bark. As you scramble down the slope, your boots collect soil smears and crumbs in which are locked the pollen and spores that rained down on this patch of woodland recently, as well as in past seasons. That soil will also contain the multitude of living things that have made the soil their home, or fragments of the dead things that lived there before.

As you crouch down to make sure of your discovery, your hair brushes past twigs and leaves hanging over the body, picking up whatever pollen, spores, and other microscopic material have fallen onto their surfaces. Your traces on the landscape—the footprints you leave behind, the hairs and fibers you shed—might be easily obscured or overlooked. But what about the imprint the landscape has left on you? What if someone was able to retrieve and identify those microscopic traces, and visualize that place, or another even further afield, from the imprint it has left on your body and clothes?

Now imagine you are the killer. What traces of the landscape where you left your victim are you carrying with you, unwittingly, wherever you go?

This is where I come in, and where my own story dovetails with the history of forensic detection. In 1994, I was an environmental archeologist at University College London. Then, things changed.

It has been nearly fifty years since I began formally studying the world of plants, though the truth is, my love affair with the natural world reaches back much further. Even as a little girl, no matter how much I read about the natural world, I always wanted to know more. There is always so much more to be had; it is still the same for me now. This is frustrating because you can never reach the summit. Nobody can. The climb is grindingly hard, and it continues forever.

I have spent much of my working life hunched over a microscope, scanning sample after sample, trying to determine the mixtures of palynomorphs—microscopic particles, including pollen grains and fungal spores—that have been stained red, embedded in jelly, and spread upon my slides. To the lay observer, what I am looking at might appear nothing more than a chaos of differently shaped blobs and splodges, but to a palynologist, one who studies pollen and other palynomorphs, these represent elements from the diverse breadth of the natural world.

Look at a pollen grain through the lenses of a high-powered microscope, and few people would fail to remark upon the strange, complex beauty of the minuscule world that was revealed to them. One pollen grain might be an orb pitted with tiny holes, while another may be shaped like a dumbbell, its wall perforated with gradations of apertures. As well as having varying kinds and combinations of holes and furrows which, themselves, might be of different size and form, the surface wall of a pollen grain might have intricate ridges of swirls, stripes or wrinkles, or networks of little columns, tracing the surface. There might be simple lumps; these may have spines, and even those spines might have spines themselves. The simplicity or complexity of form and sculpturing allows us to identify and classify these tiny products of the male cone of the conifer, or the anther of a flowering plant.

You may marvel at these tiny, often beautiful specks which are so essential to the continuation of the species. Perhaps you might even get carried away in some romantic fantasy. But, much to the chagrin of my very romantic husband, I am rather pragmatic and down to earth. I pride myself on "seeing it as it is" and try to banish any cognitive bias in the interpretation of what I see. And this is because, in my profession, these grains are much more than a stage in the life cycle of a plant or fungus. To me, they are the foundation of the stories I unravel for the police. They are the telltale signs that reveal you were not where you said you were. They are whispers telling me that you are lying or twisting truths. They are the threads that, woven together, can create a reasonable explanation of what and where, who and how. When a crime has been committed, my role is to read and present the possibilities told by the grains of pollen, the fungi, lichen, and microorganisms that have been retrieved, to try and piece together facts from the natural world.

In the past, I have described myself as a professional solver of puzzles, and the analogy is not so very far from the truth. In this profession, being accurate matters, but identifying one pollen grain or spore from another can be a taxing endeavor. One always tries to be accurate and, if there is any doubt, it is essential to use reference material of properly identified plants. Errors could contribute to someone incorrectly losing or retaining their liberty, and so long hours of my life are spent studying the infinitesimal, trying to differentiate one grain from the next. This is far from straightforward.

An ancient family such as the *Rosaceae* (the rose family) invariably has pollen grains with three furrows, three pores, and a surface with stripy whorls. The patterns of one species can merge into another so that it is difficult to be sure whether you have bramble, rose, or hawthorn, although it is fairly easy to separate this group from that containing blackthorn, plum, and cherry, where the stripy whorls are more defined and easier to see. A crime scene might be in a cherry orchard but you could never, hand on heart, claim that the pollen you have under the microscope was truly from a cherry tree because there are too few differences to differentiate from, say, blackthorn. Spores from "lower" plants such as mosses have even fewer critically distinctive features to separate them. Plants that evolved later than mosses, like the ferns and their allies, have more differentiating features than the mosses, but fewer than the conifers. In turn, the conifers have fewer than the flowering plants. This is a perplexing world with almost infinite possibilities, and yet somehow we must find a way through.

The chances are you have never encountered, perhaps never heard of, anybody in my profession. Forty years ago it did not exist. In most countries of the world, it still does not. Though I am sometimes known by other names—one sobriquet that springs to mind is "the snot lady," after a method I developed to obtain pollen grains from the nasal cavities of the dead—I think of myself first and foremost as a "forensic ecologist," one who utilizes and interprets aspects of the natural world to aid detectives in their business of solving crime. Where bodies are discovered buried in a shallow woodland grave, mummified in a suburban coal cellar, or dredged from a fenland river, I am called upon to interrogate the natural surroundings to provide clues as to what may have happened on those fateful days. Where killers have confessed to their crimes but the crime lacks a body, I am tasked with identifying the traces the natural world may have left behind on a killer's clothing, his shoes, his tools, and his car, to discover where a victim has been buried or just dumped with some cursory attempt at camouflage. Where violent or sexual assaults have taken place, I am asked to investigate the way that nature's telltale traces of pollen, fungal spores, soil, microorganisms, and more can help us point toward innocence or guilt by placing the victim or culprit in either one landscape or another. And though I was not the very first person to utilize plant and animal sciences to help the police secure a conviction, since that memorable day in 1994, it has been my life's work to pioneer the field here in the United Kingdom, pushing it in new directions and defining the protocols for best practice for those to come.

This, then, is my territory: I operate at the interface where the criminal and the natural world interact.

Because of the tediously frequent programs dealing with crime on TV, many people seem to have an acute interest, and considerable knowledge, about death. They have seen hundreds of sham dead bodies on the screen and, perhaps, have become desensitized to visions of the corpse. When in contact with death regularly, you never become truly desensitized and many TV productions seem just trivial, silly, and variously inaccurate.

There is, in my opinion, an absurd assumption among many that death is just another staging post on the long journey on which all our eternal souls are traveling. This is not something I believe. In an age not so very long ago, an age that I remember well from my churchgoing childhood, people needed beliefs like this to help them face the greatest truth of all—that our bodies are but minerals, energy, and water. That, at the end of all things, energy, the life-force, will cease to flow, and our bodies, which contain our minds and memories of all that we are, will break down into their constituents and crumble back into the great mixing bowl of natural elements, from which all living things have emerged. Most of us do not like to acknowledge, and may never have even given it a thought, that the components making up our bodies and minds, the fundamental things that we think of as who we are, once belonged to something else and that, after we are gone, they will be put to another use. But this does not depress or disturb me. For me, it is the ultimate in recycling and, therefore, reincarnation, and it will happen to us all whether we have religion or not. This is just nature, and there is more beauty in this, cold and ruthless though some might find it, than in any fanciful and impossible-to-corroborate stories of the ever-after.

The only life after death is that which is made from the constituents of our bodies, released to the world by our own deaths, so that they can be used again, and again, and again. Think of your body as a fountain where water is drawn from a reservoir. Imagine that that water spurts out in a pattern which is maintained by the pressure and shape of the fountain nozzle. The template of the fountain is your body, your life itself; but, turn the pressure off and the water will fall away, back to the reservoir. That water is analogous to the food and fluid you consume to provide energy and give you form. That form is transient, and after a brief few moments of glorious cascade, it may swirl or just dribble away, but it will inevitably join that reservoir again. If the nozzle is changed, a different template is made—a different "life" is formed. Our bodies are like the shape of the fountain, with energy and material coming in and flowing out. The "water" that makes us up will always go back to the reservoir.

No, there is no life after death—but there is always life in death. When you are alive, your body is a mass of beautifully balanced ecosystems, and so it is in death. Your dead body is a rich and vibrant paradise for microbes, a bounty for scavenging insects, birds, rodents, and other animals, some of which will come to your body to feast upon your mortal remains, and some of which will come, like the tinkers and traders exploiting a "gold rush," to prey on the scavengers themselves. And this too is of significance for a forensic ecologist—for the way a body is being broken down, the kinds of scavengers that come for it, and at what rate, can itself provide vital pieces to the puzzle of who, what, where, and how.

Maggots and carrion beetles, flesh flies and wasps; mice and rats, and carrion birds like ravens and rooks; foxes and badgers, earthworms, slugs, and snails. All of these have a part to play in the story of my work.

It is almost time to move on. But first a word about the journey ahead.

This book is not a life story, because the stories of our lives are by nature too vast and perplexing to be constrained by the pages of a single volume. This is not a textbook teaching you how to become a forensic ecologist. The scope of forensic ecology is so very wide and interdisciplinary. It involves aspects of botany, of palynology (the study of pollen, spores, and other microscopic entities), mycology (the study of fungi), bacteriology, entomology (the study of insects), parasitology, human, animal, and plant anatomy, soil and sediment science, statistics, and many other "ologies." One needs to understand the structure, lifestyles, and distributions of organisms, large and small, and their interaction with the physical and chemical environment, as well as with other organisms. Learning it is a lifetime's endeavor, and getting the right result (or establishing the most likely, as in this field there are never absolutes) is often close to a kind of intuition, a feeling built up from decades of experience of looking at the natural world holistically and using empirical science to get answers.

But nor is this a book about death.

I am not frightened of dead bodies. To me, corpses have ceased to be people; they are repositories of information where nature has left clues that we might follow. Very few times in my career have I let my guard down and been affected by the cadavers in the mortuary. The first was a twenty-two-year-old prostitute found dead in a wood, leaving three children behind. I was deeply sad for that girl, not because she was dead, but because of all she had suffered. She had been rejected by her parents at sixteen and forced to make her own way. She became insidiously controlled by a pimp who purposefully made her addicted to cocaine and then put her to work to support him and her drug habit. She bore three children, not knowing the identity of any of the fathers, but she would not give them up and her scrawny, scruffy little body bore testament to its neglect as she serviced men so that she could keep her children and cope with the rest of her existence. I cried over that girl as she lay exposed and cold on the stainless steel surface of the mortuary table, not because she was dead, but for all the struggle and misery she had suffered in her pathetic existence, while keeping steadfast in her loyalty to her children. I so admired her for that.

One of the other cases that moved me was that of the murder of a fifteen-year-old Scandinavian girl, so completely perfect as she lay there on the slab, naked in the harsh lights of the mortuary. She was killed in woodland on a lovely summer's day because of a man's frantic lust and his obsession with seeing her naked as he was kneeling in the grass masturbating. Her physical perfection moved me to deep sadness, for the life she might have had, should have had, but never would.

I have often stared at death, not only the deaths of those whose stories I have tried to piece back together but also the deaths of my own loved ones. I lost my parents, as we all do, but, before that, and sooner than I was prepared for, I lost the grandmother who half raised me; and, when I was still young myself, I was there when my daughter, not yet two years old, slipped away. My fanciful mind still sees her as a little girl in a Margaret Tarrant children's picture book—where the whole of life is portrayed as being sunny and perfect. But my pragmatic self understands my fantasy. I have been close to death myself, and I see it for what it is: oblivious and dispassionate, just another one of nature's many processes, as unfathomable as any of the rest.

Consider this book as a journey through the world I work in, and me your tour guide into that fascinating edge-land where nature and death are intertwined. Along the way, I will take you to the hedgerow in Hertfordshire where my eyes were first opened to the potentialities of plants playing a role in criminal investigation—a moment that transformed my academic view of the natural world, and the new possibilities contained within it.

I have sat at crime scenes for hours with maggot-ridden corpses, and been to the place known as the "Body Farm" in Tennessee, where cadavers are laid out to rot so that we might learn from them. We will go to the apartment in Dundee where blood-soaked carpets and cushions, thick with gray and brown growths of mold, provided vital evidence to identify the time of a murder victim's death. We will go through dense plantings of trees and across lonely moorland; to a body left on a roundabout; and then to shamanistic rituals, harnessing the hallucinogenic properties of toxic plants in the heart of southern England; so, on to the shallow graves of too many girls who went missing, never to be seen by their loved ones again. Along the way I will lead you on forays into my own history: my loves, my losses, and the narrow, little valley in Wales where I was awakened to the wonders and breadth of the natural world. If, come the end, I have left you with a little bit of the wonder I find in looking at plants, animals, and microbes, and perhaps with a new appreciation of how we, as human beings, operate within nature, not set apart from it, then I will consider my work a success.

The fact of the matter is that too few of us really understand how interconnected we are with the natural world. The vast majority of us now live in towns and suburbs, but whether we live in cities or whether we live in the remote reaches of the country, nature is everywhere. We might be the most interfering species ever to have walked or crawled on this planet, but we share it with more than a quarter of a million species of plants, 35,000 species of mammal, bird, fish, and amphibian—and, at the best current estimates, around 5 million species of fungus, and perhaps as many as 30 million different species of insect. And all of this without even mentioning the myriads of unknown microscopic species on which so much forensic ecology relies. There might be 7 billion of us on this planet, but for every one of us there are more than 200 million insects. When you think of it like this, it is perhaps no surprise to learn that nature is marking us in every step we take.

Nowadays it is fashionable to say that we live in a surveillance society, but your movements can be tracked by more than cameras. I can tell the kind of place you have been by the microscopic particles on your shoes. I can see which route you walked home, through bluebell woodland or across a garden, by the pollen on your shoes. I might tell you where you lingered with a loved one, which corner of a field you waited in, which wall you leaned on when you were waiting for your friend. And if you are one of those unlucky souls who come to me as a cadaver, by measuring the molds growing on your skin and clothes, and by the pollen and spores in your hair, clothing, and shoes, I might tell your loved ones how, where, and when you died. I can tell you who took your loved one away by the pollen embedded into his boots as he carried her off to put her in a shallow grave. By retrieving pollen, spores, and other particles from the membranes lining their nasal cavities, I can tell you if someone was buried alive, or snorted up the surface of the grave as they were being strangled. Nature leaves her clues all over us, outside and within. We all leave our marks on the environment, but the environment leaves its marks on us too, and, although she sometimes needs to be coaxed, nature will invariably give up her secrets to those of us who know where to look.

CHAPTER 2

Searching and Finding

There was a girl who went missing. In the world we live in, there are too many stories that begin like this, but one in particular lodges in my mind. Joanne Nelson vanished on Valentine's Day in 2005. By all accounts, she was bright and vivacious. She lived in Hull, East Yorkshire, and had dreams of traveling the world. She had strawberry blond hair, cut in a bob with a fringe just above the line of her eyes, and her colleagues at the local Jobcentre had no idea where she might have gone. As far as her parents knew, her boyfriend idolized her.

Of course, I knew nothing of this. The first thing I knew about the Valentine Girl was when, already missing for eleven days, the police rang and asked me to help find her.

This is often how it begins for me: the unexpected call that draws me out of bed and onto the motorway to travel anywhere the police are waiting at a crime scene. Sometimes I will be up at dawn, standing over a ditch, or in a lonely motorway turnout, looking over a body and taking samples from its remains, or, when a call comes in, I might be in my study at home, surrounded by books, papers, journals, and reference materials, with my cat on my lap, and my microscopes in the room next door, ready for action. Other times I will be in the laboratory or listening to a lecture at some scientific meeting. Familiar questions hit me one after another: Can you help us? What can you tell us, Pat? What can you show? Often the police have a limited understanding of what I can do, and what I need to be able to capture the traces nature has left behind to build up a picture of the possibilities—the might-have-been and the probably was.

This time, all the police knew for certain was that Joanne Nelson was dead. She had died eleven days previously, strangled by the hands of her lover. Her killer thought he had been wily enough and clever enough to fool the world. He appeared on camera, pleading for his girlfriend's return. He gave interviews to the press and stood alongside her unknowing parents, summoning up tears. But his tears were for his own plight and not Joanne's.

Murderers can be vain and arrogant; they often return to the sites of their killings. People say this is wickedness, driving them back to the murder scene to gloat—but perhaps they go back just to check on their handiwork, or perhaps they are compulsively drawn back to what they have done. But Joanne's murderer had no need to go back to the scene of the crime because it happened in his own home. He had strangled her in the kitchen of the house they shared, overpowering her easily after she nagged him over household chores. He had had enough; fury welled up in him and

16

he lost his temper. When a crime happens in the home, the avenues for a forensic detective to explore are often limited. Homes are full of our fingerprints and DNA and they will be coated in fibers left behind by our clothing. Joanne's house had been scoured and nothing much had been found but, thankfully, the truth had already come out.

For a time, Joanne's boyfriend had kept up the façade of innocence. He told the world she had run away. He made plaintive pleas for her to return to the family home. But the secret was too terrible to keep and, when he confided in a friend, and that friend confided in his mother, the truth came out. Paul Dyson confessed to the killing he had earlier denied. The police had their man, but there was a problem. There was no body.

Dyson could drive, but he had never gained a license. He only vaguely knew the roads of Hull, and beyond that, one road was much the same as any other. On the night he murdered Joanne, he wrapped her body in plastic and drove her as far as he could from familiar territory. He drove furtively through the night, along unfamiliar country lanes, until he found a lonely spot where he could bury her. Now, over a week later, he could recall so little of where he had been that it might have been anywhere in Yorkshire that needed less than half a tank of gas to get there. The area I had to consider was vast.

"What can you do for us, Pat?" the policeman asked me. And my question to him was the same as it always is: "Well, what exactly are you asking, and what exhibits have you got so that I can try to give you answers?"

My work often begins with things that you might think mundane, and so it was here, when the police provided me with the killer's jeans, a pair of Nikes, a pair of Reeboks, and a garden fork found at his parents' home. Paul Dyson had disposed of Joanne's body in her own Vauxhall station wagon, and this meant that evidence in the form of pollen grains, spores, or other microscopic palynomorphs could be retrieved from the vehicle. I asked for the car's passenger's-side and driver's-side footwell mats, both rubber pedal covers, the mat from the trunk, and the front spoiler from the bodywork. Exhibits such as these are my stock in trade: the shoes worn when a lover carried away his or her partner to bury the body; the material in which a still-warm corpse was wrapped; his trousers and jacket. These have been dutifully taken, logged, recorded, and sealed in evidence bags by crime scene officers.

"What can you get from things like these?" you might ask, and many policemen still do. On one level, the answer is straightforward. Edmond Locard, the French criminologist and pioneer of the forensic sciences, who lived from 1877 to 1966, is associated with the maxim "every contact leaves a trace," and this has become enshrined in forensic lore as "Locard's exchange principle." This clearly impressed Sir Arthur Conan Doyle, who once visited him in Lyon. What Locard postulated was that every time a criminal enters a crime scene, he both brings something with him that he leaves behind, and takes something of the crime scene away with him. Both of these can be used as what we call "trace evidence"—whether that is DNA, fingerprints, hairs, fibers—or the pollen and spores around which my own work is centered. They help us establish contact between people, objects, and places, and also, on occasion, provide a context for the time.

On another level, though, a case like Joanne Nelson's captures perfectly how the role of the forensic ecologist diverges from other forensic work, such as DNA analysis. I might be looking to retrieve trace evidence from the exhibits I have been given, but that is only the precursor to the main event, because what I am really searching for is an image. An image of a place that is half-imagined and half-real. What I am doing is absorbing all the information I can and using it to paint a mental picture of a place that I have never visited, and quite possibly never will. I call the image the "picture of place"—an imagined construct, but out there somewhere. This picture represents something real, a place I can summon into being by carefully considering the pollen, spores, and other microscopic matter that I retrieve from exhibits. It is the place I can see on the backs of my eyelids whenever I close my eyes. Some parts of the image are sharp, others are murky, sliding around like amoebae as more information is gleaned from the microscope. The place where you buried your lover, the place where your victim says you pinned her down and raped her, and which you say you were never near. It's the place where you picked up the telltale clues that will one day expose you, and it is how nature relays the stories that nothing else can.

And so, two pairs of tennis shoes, the foot pedals of the car, and a garden fork. These were the objects that might yield a picture of the place where poor Joanne Nelson lay.

I set to work.

My job is to give answers—or intelligence that might lead to those answers. It is work that can be long, tedious, and tiresome. For long hours I will sit hunched over my microscope before I stand up and stretch, wandering around and allowing my neck to rest. I might go straight back to the microscope because I had found something interesting and wanted to press on, or I might take a walk in the garden with my cat, or even play the piano, which stands against the wall in my study. The concentration it takes is heavy and must be sustained for long periods. Remaining focused matters more than anything else as, without this, any hope of correctly visualizing the picture of a place can vanish.

Hours can pass by. I have driven myself to exhaustion trying to decide whether the elements in a spine on a pollen grain are straight or oblique, whether the faint pattern of swirls is more characteristic of hawthorn, or some other member of the rose family. It is on decisions like this that cases can succeed or fail; a person's freedom might rest upon the difference between one identification and the next.

As I scan and count the various grains, I am building up images of plants and, from them, the habitats in which they have been growing. When an assemblage of pollen types eventually emerges from the microscope slides, it gives an idea of the vegetation at and around the crime scene; and from this I can get clues as to the acidity and wetness of the soil, whether the place is well lit or shady, and whether it is woodland and, if so, what kind. It might take me hours. It might take days or weeks, or longer. But sometimes, it all comes together neatly, and when this happens, the feeling of satisfaction is unparalleled. It can be like placing the last pieces into a jigsaw puzzle although, of course, there may be some wrong pieces and some gaps in my made-up picture. There may be pollen grains from other places on the shoes, and some of the plants at the crime scene may not be represented, but this is not a major worry because, if there are enough right puzzle pieces, the picture will be recognizable.

Joanne Nelson's murder was one of those rare cases where the picture emerged quickly and clearly; a few scans with the microscope and the essence of the picture was there. I did not have to think very carefully. They were wisps of evidence, but wisps that were almost tangible. Paul Dyson might not have known where he had been on the night he disposed of Joanne Nelson's body, but his belongings were leading the way.

I soon became certain that Joanne was lying in, or close to, commercial woodland; but I have learned, over the years, that one piece of a puzzle is barely ever enough. You must look closer and dig a little deeper, allow the pollen to reveal more of its secrets. Nor is the material you collect from a crime scene ever perfectly pristine and clear. Pollen might have disappeared or become degraded. There will be other microscopic plant and animal structures: the remains of micro-fungi, algae, plant, and animal fragments; all these may litter and confuse the view from which I am trying to pick out the critical evidence. And identifying the pollen grains is only the start. A bluebell woodland in Surrey might be similar to a bluebell woodland in Essex; Forestry Commission nurseries exist all over the country, and similar collections of trees may be planted in all of them. Worse still, in a single sample, a lone pine might yield a similar result to that of the edge of an extensive pine woodland. No, what you want—what you need—is to build up a picture, with all its contrasting and contradictory hues. Like a perfume, it might have one overriding scent but be suffused with others that can help narrow down the area we must search. That overriding scent might put you on heather moorland, in a pine forest, or somewhere along the coast, but the moors are vast, wild places, forests extend for miles across the nation, and in Britain we are never so very far away from the sea. No, what you are seeking is a combination, a specific mix of trace evidence, that can make your crime scene unique.

Now, holding the picture of place in my mind, I stretched to pick up the telephone. It was answered immediately and I was relieved to hear the calm voice of the kindly detective superintendent, Ray Higgins. His assistant officers had been frantic for information, buzzing like drones around a queen bee, but Ray was different. His gentle manner belied his keen competence, sharp intellect, and steely determination to find this little girl for her parents.

"Pat?"

I was thankful that we were not using Skype; as I continued talking with my eyes closed, I must have looked like some demented mystic. "Yes, Ray, I can see the kind of place she lies in."

I could sense the relief at the other end of the line. "She is in a Forestry Commission–type nursery." Ray became animated.

"Pat—he said there were Christmas trees there. That makes sense."

There were a few grains of spruce in the profile: spruce, the tree doomed to be sacrificed each winter festival. I was pretty sure that it was not a nursery devoted just to the Christmas market, though, because only young trees are chosen for that. To produce pollen a spruce tree must be sexually mature, about forty years old at least, and it would be very tall by that age. If Dyson had recognized Christmas trees, there may have been a considerable stand of them, possibly near the entrance to the woodland. It is an enigma that one finds so little spruce pollen right in the middle of a pure spruce woodland, or even near one, although, to the initiated, the answer is obvious. Foresters favor spruce trees at about the age of forty for felling, just as they are coming to sexual maturity; they are cut down in their prime, leaving very little pollen evidence of their former eminence in the landscape. If you do find spruce pollen, there must have been mature trees somewhere in the area.

Still with my eyes tightly shut so that nothing would distract me, I continued: "Looking at the results from the car, it seems as though it was driven up a forestry trackway, which had a dry, sandy soil, but might have had damp ditches at the edges, or in ruts. I can see that it is fairly open, but there is a lot of woodland with commercial conifers and a mixture of deciduous trees, mostly oak, beech, hazel, but with some sycamore and elm. There are damper soils because we have willow and alder. There also seems to be a lot of ivy and bramble."

My picture was tentative but, during fieldwork, and under the microscope, I had seen these plant communities many, many times before; and my results screamed "Forestry Commission." But although I could easily recognize the plant communities, one thing I had learned very early in my career was that no two places have identical vegetation. Each is unique in the patterning and density of various plant species, and although I could describe the kind of place that Dyson had visited, I needed something special and unusual if we were to find the location.

I could tell a lot more from the rather cursory results, but pinpointing the location would take a great deal of extra work. Within limits, I can fairly easily describe a picture of a place from the pollen evidence, but to find the location of the image is not quite as easy, especially if I am not familiar with the particular part of the country in the frame. I can predict the soil type and, often, the underlying geology, but in this kind of case, it is generally more efficient to ask someone who knows the local botany to use my description and do the groundwork.

I continued, "Ray, you will need to go along an open track and eventually, probably quite near the path, there will be a stand of mature birch trees. That's where you'll find her. Oh, and . . ." I stopped, because I was certain the next words I was about to say would seem the most unbelievable of all, and yet, I was certain they were true.

"She won't be buried under the ground at all."

There was a silence, and I could now sense a measure of disbelief, but he just listened while I carried on: "She's in a hollow, off the path, and will be covered over with birch twig litter."

I let that last thought of Joanne's resting place linger, because this was the image I had seen most vividly.

"How certain are you?" Ray asked.

It is a question you must always ask yourself, so you must also forgive it of others, at least if they ask politely without an accusing disbelieving challenge I so often encountered in the early years.

"Pretty certain, Ray."

There was a time when I was astounded myself at the specificity of the evidence I could produce, the detail that I could spirit up out of the images at the end of my microscope's lens. But no more. Witnesses fabricate and misremember details so that two accounts of the same moment are only exceptionally exactly the same; video and still photography capture only part of the story, leaving out the wider context and subliminally directing your thoughts; but pollen profiles can be interpreted by competent palynologists who, in addition, have extensive field experience. There are always surprises, though, not covered by the textbooks and, again, this is where experience is golden.

From all the samples I had prepared, it was clear that Joanne's boyfriend had visited a wood which had, in addition to spruce, other commercially important conifers, including pine and some western hemlock. There were also deciduous trees, with birch being utterly dominant in the profile. It was an interesting assemblage of trees and other plants, and their distribution between the various exhibits was revealing. The place obviously had an acid, dry soil, but there were damper areas too. It seemed to me that the vehicle's spoiler would give me a wider picture of the place because, as the car had been driven to the final deposition site, it would have picked up evidence all the way from the entrance of the woodland to where Joanne lay. Most likely, Dyson's feet would only have picked up evidence from the actual place where he had put the body, and this would also have been carried on his shoes to the inside of the car.

Birch dominated the profile, but pine was reasonably abundant. Oak, hazel, beech, heather, ferns, and grasses, typical of woodland and woodland edges, added to the jigsaw of an image. I was looking at residual pollen and spores; they had been produced in the previous growing season, or even before that. Hazel flowers from about December, but the others in the profile would have been produced from the later spring onward in the previous year. In other words, the evidence had been hanging around on the surface of the ground, in soil and on vegetation, since the previous season, probably even further back than that. No matter what the time of year, there is nearly always something for me to analyze and build up a picture of a place, even if that place looks unpromising to the police.

I continued to search the slides.

I retrieved nothing at all from Dyson's jeans, and mostly grass pollen from his Nikes. Obviously, he had probably not worn these when the offense was committed. But when I looked at the slides from the Reeboks, the foot pedals, and the car spoiler, the place leaped out at me. It was already clear to me that we were dealing with a woodland dominated by commercially grown conifers, but now, as more slides were scrutinized, other pollen types started to be revealed. The mixture of palynomorphs made perfect sense. Nursery owners often include native deciduous trees, including birch, at the edges of their nurseries to mask the dense, boring monoculture of conifers. Birch cannot stand the shade, it grows well in poor soils, and its life expectancy is about the same length of time that it takes conifers to reach a size suitable for cropping. One can see why this "weed" tree is certainly a favorite of nursery owners.

The richness of the material retrieved from the spoiler of the car suggested to me that Dyson had driven the car deep into the site, looking for a likely place to bury Joanne. The insides of the car revealed a miniature mystery nestled inside the bigger one. Even though Dyson's Reeboks had been thick with pollen from the woodland itself, none of this seemed to have transferred to the driver's floor mat—which was unusually clean, betraying only a couple of grains of pine pollen and a single grain of heather. Was Dyson sufficiently aware of forensic science that he had brushed, vacuum cleaned, or even scrubbed the floor mat? It certainly seemed so. Yet the mat on the passenger side revealed the same rich assemblage as his Reeboks had, and images of how this came about flashed into my mind: to lift something big out of a car, you sometimes have to step into one of the footwells to brace yourself there. I could imagine Dyson, readying himself to lift Joanne out of the car and, grappling with the weight of her's body, planting his foot down on the passenger's-side mat to get better leverage.

I had established that we were looking for a commercial forest but I needed something to pinpoint its location and the kind of place within the woodland where Joanne might have been placed.

Realization came by virtue of the garden fork.

There was a lot of birch pollen on Dyson's footwear and where he had transferred it to the passenger's-side mat of his car. But, when I examined the sample from the garden fork, I was staggered. There was birch pollen up its shaft, birch pollen on its handle, but on the tines it was thick to the exclusion of almost everything else except a few grains of pollen from typical garden plants.

An image of a place suddenly forms in my head. It is not as magical as it might seem. It is intuition built upon all my years of slogging through basic subjects, trekking, and working in the field, and continuously accumulating reams of knowledge about the natural world around us. All this is stored and processed in a re-markable supercomputer, the human brain.

This time I saw it clearly: how Dyson had driven Joanne's Vauxhall station wagon along a Forestry Commission track of dry, sandy soil; how the commercial conifers had risen up starkly around him until he found a place where the trees were not so dense, and eventually he came to a stand of birch. This seemed the perfect place. Digging a grave is hard work, and digging a grave with a garden fork is impossible. A fork is no good for digging a grave but it certainly is good for scraping.

If you spend any time wandering through woodland and com-mercial forests which are often at the edges of heathland and moor-land, you notice the dense buildup of twiggy litter. It collects in hollows, and you can get caught out by thinking there is solid surface until you stumble into a hole. The floors of commercial forests are riddled with undulations, where clumsy furrows made by forestry workers create hollows and hummocks, and are mementoes of past planting and felling. Why bury a body when you could find a suitable hollow and just scrape litter over it to mask its presence?

I suspected that he had carried Joanne's limp body, placed her in a convenient hollow not more than 100 meters or so from the track. Behavioral profilers have worked out that about 100 meters is the limit of someone being willing and able to lug the dead weight of a corpse. I imagined him stumbling, his Reeboks sinking into twiggy litter, he would have panicked, realizing that it was too difficult to bury her. But he could hide her. I thought he could have rolled her into one of those hollows and, using the garden fork, scraped the birch twig litter over her, masking her from view. Of course, she had to be in a hollow so that sufficient litter could cover her. If she had been left on a flat area, there would have been a body-shaped mound of litter, and this would have been very obvious.

He must have thought it would be a very long time before she was found. In that bitter winter, close to Valentine's Day, not much forestry work would have been going on. It was very cold for bluebottles to be searching for sites to lay their eggs, and the smell of death would be delayed. This meant that foxes and badgers may not have found her as readily as they do bodies buried in high summer. She might have remained undiscovered for ages— perhaps forever.

As he was scraping the twigs over her with the fork, he was blissfully unaware that he was also coating the fork's tines with the birch pollen that would later help bring Joanne Nelson home, and lead him to court.

"But where is this place, Pat?" Not close to Hull, that is for sure. It has to be a nursery that included western hemlock, all the plants in the assemblage, and a specific fern that, at first, did not ring too many bells for me because it is so common over so much of the country. This is *Polypodium*, the polypody fern. It is incredibly common in the south and west, and even in Surrey where I live, but in this part of Yorkshire it is decidedly uncommon, bordering on rare. What a piece of luck. The distribution of this fern could be checked in the maps of the Botanical Society of Britain and Ireland, something I do on a regular basis. What was so interesting was that the polypody fern had been present in the area in the past, but was now absent from the area altogether. The historical records showed where it had grown before, and this enabled huge areas of terrain to be eliminated. There were three possible woodlands that had western hemlock, but only two of these had had polypody fern in the past.

Dyson had already said that he remembered a metal gate with lots of empty bottles lying to one side of the entrance. The sergeant in the case was brimming over with enthusiasm and could not help calling in various experts, sometimes inappropriately and irritatingly. However, a local botanist in Hull looked up the record maps and found the historical distribution of polypody fern before I had a chance to suggest it myself, but that did not matter. We just had to find this young woman for her parents' sake. Not for Joanne's. She was gone, but her grieving and fearful family were desperate to have her back, and Ray Higgins was determined to bring her back to them.

The police were beside themselves with the new information, and they set off with Joanne's murderer cuffed in the back of a police car to scour the winding roads of that landscape, looking for a metal gate and bottles. They were just too keyed up to wait, and they drove along miles of roads until, weary and dispirited, their eureka moment came: here was a pale metal gate, and the bottles were still there. I would love to know how they felt but I expect it was excitement mixed with relief.

They found Joanne's body quite quickly, and later said they were taken aback at the accuracy of my description. She was down a slope, just off the track, in a hollow, beneath birch trees and covered with birch twigs. So, years of slogging at basic subjects, trekking and working in the field, putting several and several together, binding it with common sense, and having the courage to speak the vision, paid off.

I was not there on the day they actually found Joanne Nelson's body and finally confirmed for her anguished family what they must have known all along, but I have often imagined how it must have felt to be one of those police officers who were initially skeptical of my vision. Long after Dyson had been convicted, Ray took me to where Joanne had lain. I had been taken aback at the accuracy of my description of a hedgerow and field in my very first case, which I will come back to later, but I was just as shocked as I walked with Ray through that fateful gate. Here was the sandy path used by forestry vehicles. It was deeply rutted and sparse heather sprouted both on the top of the ruts and along the tops of the gouged edges. Common heather tolerates damp conditions but does not like standing water; the hollows were certainly damp, however, and a shallow, stagnant ditch rang along the path edge for some distance. There was an open area to the left dominated by bracken ferns and, as far as the eye could see, dense trees stretched from the right side of the path and straight ahead. The tree plantings reflected what I had found on the microscope slides: there was a lot of pine, western hemlock, and spruce.

There is one facet of the Joanne Nelson case that always helps me explain the way forensic ecologists can work. I had come to know the type of landscape we were searching in. I knew its trees, and so the nature of the soil, and the non-tree plants on the ground that could be expected to be growing with them. I felt as if, in my mind's eye, I could follow Paul Dyson down the track to where he had hidden the body. But Yorkshire covers a huge area and there could be many places that might have fitted the template I had built up. After all, there are forests with stands of birch trees across the county. The police could perhaps have scoured them one by one, if they had the resources to do so—but nature had given us one final clue to point the way. For there, in the microscopic remains retrieved, was the pollen of western hemlock and the spores of polypody fern.

Even after all these years, I never fail to be impressed how pollen profiles can give so much information, thoughts, conjecture, and visions, all based on scraps of evidence that, although tangible to me, must have seemed like superstition or magic to others.

In recent years, some geologists have been trying to help in cases like these. One in particular, twenty years ago, worked on cases, convincing police officers that trace evidence provided by mineral particles in the soil was infallible. He became known by the size of his invoices, which were invariably inversely proportionate to the amount of information he could provide. He cost various police forces a great deal until he was eventually discredited and, even with his protocol of taking literally thousands of samples, he could not get the resolution that plants can provide from a few simple samples, a microscope, and a dash of medicated shampoo used to wash the exhibits. In one case he came up with the same location as me, but at the cost of analyzing over a thousand samples collected along a transect line over 50 miles in length, whereas I reached the same result with just four samples from a spade and some observational fieldwork. If you know the ecology of your plants, you can predict the kind of soils they will be growing in and, therefore, the geology underneath them.

There is too much superstition around. But I do not do magic. This is science. Twenty years earlier, Joanne's body might never have been discovered until her scattered bones were stumbled upon by a forestry worker, or chanced upon by a solitary dog walker. But with the science I have pioneered, we were able to detect the place her murderer had visited by the microscopic traces left behind by nature. Whether we are murderers or not, we leave our paths behind us—and a person who understands landscape, pollen, other palynomorphs, fungi, and soils can follow them.

I received a letter from Joanne's parents thanking me for bringing their girl home. I must say that, until then, I had thought of Joanne as more of a puzzle and was too engaged in finding information to think about her as a person. In a world where you are faced with some of the worst things that human beings can do to one another on a daily basis, it is all too easy to become desensitized, to lose yourself in the intellectual challenge and park the human cost. Sitting in my study with my beloved cat upon my lap, reading Joanne's mother's words, something changed in my perception. Joanne Nelson was not only a puzzle that needed to be unpicked. She was not only a challenge for me to confront with the years of experience I had acquired. She really had been a living person, with loves, hopes, fears, and ambitions. With the words of her mother came that realization, and with it a rush of emotion that I rarely feel. And this, more than the intellectual challenge, more than the pride I have always taken in advancing the science of forensic ecology, is why I do what I do. People's feelings matter.

CHAPTER 3

_____

Proxies of the Past

It is time, perhaps, to go back to the beginning.

I did not mean to become what I have become—and this, of course, is how all the best stories start. I was already in my early fifties when I got the telephone call that would change the course of my life, drawing me into the world of forensic investigation. By then, my career was already what you might call wide and varied. I had started my working life training as a medical laboratory technician at Charing Cross Hospital, and was involved in the second renal dialysis unit in Britain. I was certainly used to dealing with blood and excrement and having to put up with vile odors; it was part of the job. Eventually, I became involved in research projects, and this meant dealing with laboratory animals. I came to love the rats—white, pink-nosed, twitching little things. They were curious and loved being cuddled and tickled. I loved the animals and hated the research program, and I decided emphatically that medical research was not going to be my career.

My boyfriend, whom I later married, thought I should do something more "ladylike" than working with rats, analyzing urine, feces, and blood. But what, I wondered, did "ladylike" really mean? Perhaps the full-time business and secretarial course I had just seen advertised was what I needed, so I applied and got a full-time, funded place. The course proved to be quite hard. It was new and the college employed part-time practicing professionals to lecture in the core subjects—law, economics, psychology, and English. In addition, there was great pressure to excel at typewriting and Pitman's shorthand (the most logical, yet flexible, and wonderful system I have ever encountered). We also had to learn just about everything needed to run a high-powered office. After all the studying I have done in my life, I look back on that course as being truly outstanding. I loved the challenge and achieved a distinction for the diploma. Come the end of the course, all the students were entered into the examinations run by the London Chamber of Commerce—these were competitive and international, and where secretarial and business skills were tested and scrutinized. I was staggered when I came first, but quite enjoyed all the publicity and fun that went with it, including being presented with an award at Mansion House by Lord Luke.

After I graduated, my first job was working in Knightsbridge at the head office of Coca-Cola. New recruits put on pounds because the product was on tap, and my first encounter with corporate loyalty was having to be seen with a glass of the sugary, acidic concoction on one's desk at all times. I thought it was a ridiculous job, working for egotistical, dark-suited nobodies, all engaged in selling something that sold itself anyway, so I quickly moved on to work for a huge, prestigious building company. It was a demanding and responsible job, and it was fun reading about the technical side of the building of monumental projects like London Bridge and Drax Power Station but, after some years, even that job grew wearisome. It was just not satisfying and I was bored. There was too much routine and too little opportunity to learn things that fascinate: I needed new challenges. I was like a pony pushing at the field boundary, wanting to know and sample what was on the other side. I wanted to learn what was out there.

What I did next led me into one of the happiest phases of my life. I went to read botany at King's College London. It had taken me until I was in my late twenties to find my true niche, and, after all the things I had tried, I felt I had arrived. I was ten years older than most of the other undergraduates, but no one seemed to notice, and there were certainly no barriers between us; we just mixed in together. By this time, I was married and running a home, but I joined in many of the usual student activities. I was elected as president of the Biological Society, and my particular friend, Myra O'Donnell (a brilliant and highly organized zoologist), and I spent every Saturday morning in the fencing class in the gym. This was situated in the bowels of the Strand building.

Our tutor was a dashing Hungarian of a certain age who, between the flamboyant sweeping back of his hair, poked us in the ribs with his foil until we learned how to guard and riposte. Because I got so out of breath, he kindly allowed me to take my bronze medal exam in two parts, thus allowing me to keep the fantasy that I could fence. Outside the gym, the corridor had very old paving slabs and was lined with lockers dating from Georgian times, each bearing gold-painted, decorative numbers. Myra and I used to fight all along the corridor, up the steps leading to the main foyer, and back again. One Saturday I said, "Do you realize that we have created a choreography and that neither of us can win?" We fell over laughing, but continued our ritualistic Saturday routines.

I look back on my time at King's as being magical, exploring as many academic topics as I could fit into my schedule! Even though I was supposed to be a botanist, I grasped every opportunity to learn ecology, geology, microbiology, zoology, parasitology, biogeography, and anything that gave me a greater understanding of the natural world. It was a joy to spend hours in the library being surprised by texts that today's students—who rely extensively on electronic information—would never encounter. It was a traditional education, attending small tutorials, taking notes, writing essays, browsing the library, carrying out research projects, and enjoying field trips in many and diverse habitats. I learned the secrets of so many facets of the natural world, from the nervous system of a lizard to the structure of grasses. It was the making of me and I loved it.

Eventually, I became a lecturer in microbial ecology at King's. At first I thoroughly enjoyed the challenge of "being on the other side of the fence." It was rewarding to help students enrich themselves with knowledge of the natural world. I felt I was passing on something very special, knowledge that should not be lost—but eventually, my heavy teaching load, the grinding routine of writing and delivering lectures, setting and marking essays and examinations, and attending meetings got me down. So, after eighteen years in one of the happiest places I have ever been privileged to work, I applied for a job at University College London, in the Institute of Archeology. I wanted to spend most of my time on research rather than on teaching.

The botany department at King's was small, the staff were

jolly and we had lots of fun; and any occasion of note was marked with a party where everyone, from the first-year students to the professorial staff, attended. At UCL, it was very different. Here, apart from the occasional seminar, I was fully engaged in research and the atmosphere was utterly different. I had gained a new title that I was not sure I deserved. I had become an "environmental archeologist." At King's, lunch and teatime breaks were full of laughter and intellectual discussion, and I looked forward to walking through the front door to the department each day. At UCL, people kept to themselves and stayed in their rooms; it was difficult to get to know any of them. But the nature of the work made up for the lack of social life, and I soon made chums with other environmental archeologists all over the UK.

It was a most fascinating time, analyzing sediments from archeological sites and their surroundings, trying to ascertain the variation in ancient landscapes, the kinds of crops and husbandry carried out by prehistoric peoples. It meant I spent long weeks tramping around archeological sites all over the country, taking soil and sediment samples from buried surfaces, pits, and ditches back to the laboratory, and carrying out the lengthy and dangerous chemical processes involved in retrieving the organic particulates from the cores and monoliths of deposit. After analyzing sample after sample from site after site, from the Paleolithic period to medieval, I gradually realized the potential and limitations of the techniques we were using. I spent my life reconstructing environments from organic particles, pollen grains, and spores retrieved from the archeological features. The role of the environmental archeologist is to give color, life, and meaning to the settlements unearthed by the diggers.

I analyzed sites as varied as a fort on Hadrian's Wall, and a deep bog to its north, an inn excavated from the volcanic ash in Pompeii, and even the multiphase site under Terminal 5 at Heathrow Airport. My work at Heathrow revealed a most wonderful Bronze Age landscape which must have been a rural idyll four thousand years ago, with long, pretty hedgerows dividing up fields that variously held cattle and sheep, or were planted with cereals. It was like looking through very thick lenses so that although we could gain a good idea of past land use, we could only check our interpretations against modern ethnographic examples. There was no way of knowing whether our interpretations were truly accurate. There were many rewards, though, and a major one was working on sites with other environmental archeologists. I analyzed pollen and spores (palynomorphs). Peter Murphy, a very special friend of mine working at the University of East Anglia, specialized in seeds, other bits of plant you could see with your naked eye (macrofossils), and mollusk shells; others analyzed animal bone or human remains; and another special friend, Richard Macphail, who was with me at the Institute, was (and still is) a soil micromorphologist. He embedded soils in resin, cut thin sections from the blocks, and found clues to previous human activity directly under the microscope. I was nearly as interested in his work as I was my own. Just imagine being able to see a section of soil under the surface—all its minerals and organisms suspended as if in "aspic"—being able to view a microcosm in its hidden reality. These were the soils which supported the plants, animals, and people that others of us "brought to life" in building up realistic pictures of the past.

We had many happy meetings and much fun, on sites and at conferences, and collectively we created pictures of what had happened through many phases of ancient occupation and development. When you visit a museum and you see the reconstruction of a

Roman farm, a Saxon village, or a Stone Age hut, you can thank the band of environmental and other archeological specialists who have done meticulous analysis to bring you the story. To my mind, without them archeology would be as dry as dust—flints, pottery, stones, and metal, with an occasional thrill of a bone, jewel, or carved stone. Essentially, the archeologist carries out the excavation and retrieves those wonderful things from the ground in a meticulous way, but it is the specialists in the multitude of disciplines—metallurgy, pottery, entomology, botany, osteology, and micromorphology of ancient and buried soils—who bring everything alive. Not many people know that.

Pollen and spores, and many other microscopic things in my samples, are proxies of the past. If the activity of bacteria and fungi is suppressed by lack of oxygen, or by acidity, pollen can remain preserved for thousands of years. We could not miss or ignore a single one, because even insignificant, tiny particles might be informative. After the processing, samples were permanently mounted onto microscope slides, and then the grindingly hard, real work started—sitting at the microscope for hour after hour, scanning transect after transect across the slide in strict sequence, resisting any attempt to miss a single field of view in case something important was overlooked. I was fascinated by reconstructing ancient environments and it was marvelous being with colleagues who worked with other classes of evidence, our work all being combined in the final report to paint pictures of the past. I was fairly happy with my lot, so when the telephone rang that day, I did not expect it to begin a new chapter in the story of my life.

The voice on the end of the line had a thick Glaswegian accent and belonged to an officer in Hertfordshire Constabulary.

"Are you Pat Wiltshire?" he asked. "We've been given your

name by Kew. They weren't able to help us . . ." Here he paused, as if to let something sink in, ". . . but they said you could."

Only a few moments ago, I had been somewhere in the Neolithic period, building up an image of our native forests, as they were beaten back and burned down by these first farmers. Now, wrenched into the present again, I hesitated.

"Oooh, right," I said.

I was intrigued—I had never been contacted by the police before.

"What's involved?"

"You're a . . . polyologist?"

"Not quite," I said, doing my best imitation of patience at this common error. "I am a palynologist."

Palynology. Quite literally "the study of dust"—or, to put it more helpfully, the study of pollen, spores, all the other microscopic palynomorphs, and particulates that we can collect from the air, from water, or from sedimentary deposits, some soils, and vegetation. To become a palynologist had not been a grand plan of mine either, but it was the road along which life had funneled me and I was quite content. There was much more freedom than when having to be responsible for students.

The detective on the other end of the line was still waiting for an answer.

"Why do you need a palynologist?" I asked.

The voice came back bluntly: "We've got a murder." I nearly laughed because, in his Scottishness, he rolled his Rs very strongly and "murder" became "merrrderrr." All this was so improbable that it could have been from a play in the West End.

"A murder? How can I help you?"

"We have a body and we have a car."

I have often looked back on this conversation as a turning point in my life. The fact of the matter is that at the first mention of the word "murder," I had become more than intrigued. When you work in laboratories day in and day out, sometimes an intrusion from the outside world is very welcome. Apart from members of my own family I had never seen a dead body, although at King's I was responsible for teaching about decomposition and breakdown of materials after death. This needed an understanding of the roles played by micro-animals, bacteria, and fungi in breaking down bodies, whether they were of dead birds or trees. Was a dead human such a big step? In an academic sense, perhaps not, but in every other way, this was a leap into the unknown, and I was not prepared for it.

As I listened, the detective outlined everything he thought I needed to know. A body had been found in a field ditch, somewhere in rural Hertfordshire, but it seemed that this had been an accidental killing.

"We are dealing with Chinese Triad crime."

That was something one only ever heard of on TV and never really believed existed—in the realms of Sherlock Holmes. But one is led to believe that the Triads are vicious outfits whose activities have serious consequences. On this occasion they had not meant to kill their victim; they had abducted him on his wedding day, not from his wife's bed but from that of a prostitute. I have to say I was quite stunned to hear that particular snippet of information. I had never come across anything like it before. I live in a place where we do not even get graffiti, although I suppose a bicycle occasionally gets stolen from the station.

The gang had hog-tied him and put him in the back of a van,

meaning only to teach him a brutal lesson. He had been engaged by the gang for a money-laundering venture, buying and selling property, but he had made the mistake of siphoning off cash for his own use. He was a big man and being thrown onto his front while hog-tied resulted in his heart and lungs failing: he died of asphyxiation under his own weight.

When I first spoke to them, all the police had was the car that had accompanied the van when they dumped the body, but the van had been disposed of very quickly. The gang must have panicked—they had decided to dump the body in some remote place in Wales, but their sense of direction was somewhat flawed. To get to Wales from London, one needs to go west along the M4 motorway, but they found themselves traveling up the A10 northward in Hertfordshire. In the dark and being disoriented, they must have been relieved to find a side road off the A10 which seemed to lead to an isolated field.

Then came mistake number two. They put the body in the field ditch and tried, stupidly, to obliterate the victim's identity by pouring accelerant over his corpse and setting fire to it. Left alone, the body might never have been discovered except by flies and scavengers—rats, birds, foxes, and badgers. Scrub would have grown up around it and long grasses hidden it from view. Earthworms, slugs, snails, beetles, and ants would soon have moved in and, before a season had passed, especially if it had been warm, a body would have barely remained. Any bones, picked clean, would eventually be buried by busy earthworms—they will bury anything placed on a surface if left for long enough. Darwin had neatly demonstrated this by placing paving slabs on his own lawn. But by setting the body alight, this man's killers had created a beacon in the dark which was still smoking the next day. It was the rising smoke that had caught the attentions of the farmer and, after him, the police.

"We've got them in custody," said the detective, whom I now know as Bill Bryden MBE. "We've got their car. We're certain it's them. But . . . we need to prove it." The detective paused. "And then the boss thought . . . maize pollen."

His boss turned out to be one of the most engaging people I had ever met. He was Paul Dockley, a young and intelligent assistant chief constable. I had never had anything to do with the police before, and I had already met two great chaps in Bill and Paul, who are my friends to this day. They have always been very supportive of my work and, indeed, of forensic ecology.

"Maize pollen?"

"They had to drive into the field to leave him in that ditch, and the farmer told us that it was usually planted with maize. If they drove through the field, the governor thought that maize pollen could be on the car. That's where you come in. We need somebody who could tell us for sure. Did this car go into that field?"

It was a new idea, he said. It was not something any police force had thought about before. Well, I had thought about it in a vague sort of way; the odd article in popular magazines had made me think along these lines, but I never imagined I would be presented with anything like this. Why maize pollen had popped into his boss's head, he was not quite sure; I had not even seen the car but already I knew that any chances of success were infinitesimally small. The calendar on the wall read May, and that meant we were at least six weeks away from the maize flowering period in southern England. Not only that, farmers plow and apply fertilizer to arable fields so that the soils become nutrient-enriched and aerated, and these conditions enhance microbial activity. Farmed fields, particularly in southern England, are paradises for fungi, bacteria, and the rest of the microbiota, so that organic material readily decomposes. No, I did not think that pollen and spores would have been preserved in such a typically managed field. And yet . . .

"Who knows?" I began. "There may be micro-sites down there. Corners where something's survived."

"Does it sound," the voice went on, "like something you'd be able to do?"

"Well, I can try, but I have to warn you that I might not find anything." I went on to describe the potential problem with farmed soils.

The thought of a dead body did not much bother me. I could vividly remember having my arms full with a cloth-wrapped dismembered leg as I walked down the corridor of Charing Cross Hospital. It was destined for an infusion experiment in the research laboratory. A body is just a body. It is flesh, blood, and bone. No, what bothered me was that this was the great unknown. Police work was a different world, and not one I had ever thought to visit. I had no idea about forensic protocols; I had never heard the term "trace evidence," nor any of the other terms, acronyms, and phrases I would soon discover. I had, in my day-to-day life, become practiced at imagining the landscapes of the past—but to contemplate the landscapes of the present, to search for something that had been left behind, to invent procedures where none had been invented before? As I listened to the detective's breathing on the other end of the line, all of that seemed an entirely new frontier. This was a bit like *Star Trek*—"To boldly go . . . !"

Then I thought: You keep asking yourself *why* but why not? You've done it before. You've worked in laboratories and hospitals, reinvented yourself as a high-flying secretary in the building industry, reinvented yourself again as a microbiologist and yet again as a palynologist. And isn't that part of science? To be curious and try? You never planned your life before. Why not embrace the opportunity now?

If it did not work, it did not work. Nothing ventured meant nothing gained—but it would not matter to me. There was always my work in archeology, which I really enjoyed. There were always features from the past, waiting to be excavated and rediscovered. So, for the rest of the day, I thought very little about the murdered man and Bill's Glaswegian voice on the other end of the line. It sounded like an interesting exercise, but really, that was all. How was I to know it would dictate the direction of the second half of my life?

The car used for dumping the victim in the field ditch was waiting in the police garage, looking just like any other old car. Its wheel wells were spattered with dirt; there was general grime along the bottoms of the doors, with a few superficial soil smears. The police garage attendant who led me in, flicking on the overhead lights to illuminate the waiting vehicle, did not look particularly impressed. "I don't know what they brought you down for," he said, barely concealing his scorn. I eventually came to expect this kind of dismissive sarcasm in the early years. "There'll be pollen all over the thing. That car's been up and down and all over. Just look at the state of it . . ."

I crouched down on one side of the car, and then on the other. The outside of the car looked as if it ought to have been a bounty of information—but how to extract it, and where to begin, I had absolutely no idea.

I had already asked the police to get me samples of surface soil from the field where the body had been found—from the wheel tracks. I had put it through the usual process but, as expected, the microscope revealed only an occasional speck of residual cellulose, stained bright red by my safranine dye. The rest was what I called background "grot," with an occasional fragment of a pollen grain, decayed beyond recognition. My prediction of thorough decomposition of organic remains in that soil was correct. Looking at the soil traces in the treads of the tires, the accumulated tidemarks of black, silty deposits in the wheel wells—and, inside, the vague dusty footprints on the floor mats—I wondered if I would find anything at all. But the garage attendant was squinting at me, expecting me to find nothing, and this galvanized me more than ever. It was impossible to do anything under those conditions, and I was pretty fed up with the officer's attitude, so I chose those parts of the vehicle I thought might be most productive and instructed the officers to send them to our laboratory.

So much has been on a steep learning curve. The chassis elements of vehicles vary greatly, but I now know the most likely nooks and crannies that might collect relevant evidence. Back then I knew nothing—I had never even seen the underneath of any motorized vehicle, certainly not at first hand with my face about five centimeters from the oily, grimy metal of the various pipes and struts. I soon came to realize that I would just have to do my best and, by trial and error, find the most efficient way to sample these things. I was used to scrubbing the dirt from various artifacts to find out what they had contained. Could this be so dissimilar? So, I just used my common sense; I started with the most easily removed items and asked that they be brought to me—footwell mats, pedals, bumper, air filters, and radiator. Initially, I ignored the wheels because they could have picked up material from a multitude of places. On the other hand, the inside of the car would contain mostly material that was transferred to it from people's feet, and the objects they carried in it. Simple logic guided me and, in any case, if I was wrong, the rest of the car would still be in the garage and could be resampled.

I was glad to get away from the sarcastic and downright rude police garage attendant. It took me an age to scrub and wash the various items meticulously, and to sieve and decant the silty washings so that they could be centrifuged down to concentrated pellets. The worst item was the radiator, which yielded great clumps of insects. I saved these and asked a colleague to have a look at them but I did the rest.

There was pollen everywhere. The attendant who poured such scorn on the endeavor had been right about one thing: the body of the car was a veritable botanic garden of different pollen types, and they had obviously come from different sources. I subjected the centrifuged pellets to standard processing just as if they were archeological samples. This involved using a sequence of very strong and noxious acids to remove the background matrix of the soil— the quartz (sand), clay, cellulose, lignin, and humic acids. In the best possible scenario, only the various palynomorphs would remain.

As amazing as it sounds, the outer coats of such palynomorphs, pollen grains, spores, fungal remains, insects, and crustaceans all contain incredibly resistant polymers that can withstand the vicious treatments involved in the methods. These polymers are sporopollenin in the case of plants and chitin in the case of fungi and animals. The method is so dangerous that one must never be left alone in the laboratory when engaged in this work, and layers of protective clothing, gloves, and a mask must be worn. No one is allowed to enter the room when the processing is in progress, and every precaution is made to eliminate contamination by stray airborne pollen grains. Slide traps are also placed around the laboratory, on windowsills, on the surface in the fume cupboard, and randomly elsewhere. In this way, it is possible to check for contaminants floating around in the air. I also ran blanks to test the various reagents to make sure that they were not contaminated in any way.

Once the background matrix of the soil had been removed, the pollen, spores, and other organic remnants that had miraculously survived were stained and embedded in jelly. The jelly from each sample was then spread thinly on a glass slide and allowed to set. Only now could the really hard work begin. I had never examined objects such as vehicles, clothing, footwear, or anything else so modern and mundane. But I was enthralled. Every sample from the car contained abundant pollen, spores, and fragments of insects, as well as microscopic entities that eluded immediate identification.

The radiator mesh at the front of the car had drawn in everything to which the car had been exposed. Here was a hotchpotch of remains of organisms that indicated the rural and the urban, of farmland and woodland. Who knew how long it had coated the inside of the mesh? The tires, too, were so palyniferous that they certainly represented more than one place—probably hundreds of places—so too the chassis, where shreds of material, bits of plant, clods of earth, and puddle water had all left their impression. I had managed to extract a rich mosaic of microscopic matter that was diverse and well preserved, yet proved exactly . . . nothing. The total information was rich and bountiful but so muddled that it was worthless. Yet, as the work progressed slowly, I realized that some parts of the car were giving more specific results. The tire treads themselves had picked up dirt from everywhere the car had been driven—but the inside wall of the tire represented more singular landscapes, with much less pollen finding its way into these hidden corners. I began to see that different parts of the car accumulated different material. The differences were not large, but we work in a microscopic world. Small differences matter. One thing that was very noticeable was the unexpected dominance of tree pollen in samples from the outside body of the vehicle.

Then I started investigating the insides of the car, enjoying myself as I began hitting my stride—and everything changed.

I had not been expecting it—but then, I had so little idea just what to expect. The inside of the car was much cleaner and free of grime, at least the kind of dirt visible to the naked eye. I analyzed the fabric from the seats, the air filters, the window frames, every nook and cranny inside that car, and the results were just not interesting at all, but what leaped out at me was that the profile of the foot pedals, and the mats beneath them, matched. It was not a perfect match. Nothing ever is in this kind of study. But their profiles both reflected the same kind of place. They both had pollen of dogwood, dog rose, oak, hawthorn, bramble, field maple, ivy—and lots of *Prunus*-type pollen. Looking at the rest of the assemblage making up this plant community, I was certain that the *Prunus*-type was blackthorn, the very common plant that generously gives us its little plumlike fruit which makes lovely sloe gin. What interested me, too, was the preponderance of pollen from weeds normally found at the edges of arable fields. In archeological contexts they are regarded as indicators of past crop-growing—black nightshade, poppy, white dead-nettle, stinging nettle, woundwort, docks, goosefoot, and grasses, to name a few. I also found some cereal pollen. Of course, cereals are grasses and their pollen grains are very similar to each other except for size. These grains were obviously not of any grass because they were too big; they were not maize because they were far too small; and they were definitely not rye, because they were round rather than a tapering, rounded oblong shape. They were probably of wheat or barley. A picture of this place was beginning to form in my head. I had been told that the car had driven into a field that was used for growing maize. I did not find any maize pollen, but then I had not really expected it. The way that field soil had been enriched and aerated would have promoted its disappearance by the late winter or early spring. Such treatment would enhance the activity of decomposer microbes, and pollen would just disappear. But the soils at the edges of the field would not have received regular doses of fertilizers and pesticides, or been aerated by plowing. Importantly, all this meant that microbial activity would be much reduced and pollen and spores might be better preserved. In any case, with such a profuse growth of herbs, anyone approaching the ditch could not have failed to step on them. These would deposit their own pollen on shoes, and that of the shrubs and trees of the hedge, as well as being brought by the air from farther afield. Even pollen and spores from previous years could remain impacted on leaves and stems, and be present in the ditch itself.

Going through samples and recording your findings is a painstaking, tedious business, but moments like this—when an image slowly forms in front of your eyes—give great rewards. This was an assemblage of pollen I had seen many times before. In fact, it was a typical archeological assemblage, the kind that has been in Britain for many thousands of years since the first farmers started cultivating crops, and field margins developed.

There was no maize in that field, nor its pollen, but I was

looking at the vegetation that had been growing around the crop in previous years, and probably would for many to come. This was evidence that the occupants of that vehicle had stepped on the vegetation growing at the edge of an arable field, that was bordered by a species-rich hedge. The number of tree and shrub pollen types I found probably meant that it was an ancient hedge, and most likely it would have been growing on a bank thrown up by the digging of a ditch in the distant past. To all intents and purposes, we were looking at an archeological feature. It was coming together.

I picked up the telephone.

The sun was slanting through the Hertfordshire trees as the detective's car came to a halt, and I clambered out of the backseat. I had jumped at the chance to come out here and see it for myself and here was the hedgerow, stretching far along the edge of the field. There, on the other side of the road, at least 200 meters away, was a field planted with wheat—that probably explained the cereal pollen I had found, although I was rather surprised that it had traveled as far as it had. Experiments by some had shown that it travels only a matter of a few meters from a crop edge—so here was another anomaly in the standard literature.

Field margins up against hedges may be wasted to the farmer but not to wildlife. They are teeming habitats. They are the homes for hundreds of different species: plants, insects, birds, and other animals. And the one whose brambles, nettles, and herb-rich bank stretched away from me now was the place where a man had been dumped and set alight; in the hope that his body would never be found.

The hedge was of various heights, depending on the shrubs

growing within it. What a bountiful place it was, with so many different species, each producing specific pollen, leaving traces on whoever touched them. What a contrast to the sterile, bare soil of the field. If there had been a field of maize when the vehicle drove into it, the evidence would have been so easy to interpret. Yes, the field held maize and yes, the vehicle and its occupants would have been swamped with its pollen. Getting evidence like that is very rare, though, and this was certainly not the case here. A place as rich and varied as this left so many more markers but, if I had not analyzed so many ancient arable field ditches, I might not have put two and two together and realized its significance.

I have thought about hedgerows a lot in my life. In archeology we might find a certain assemblage of pollen grains indicating a hedgerow and interpret the results to tell the archeologist that the land had been farmed here in times past—and in that way begin building up an image of how our landscapes might once have been. Now that I stood here, another thought leaped out. What made one part of this hedgerow different from the next? There are hedgerows in Britain several millennia old. The oldest hedgerows hark back thousands of years, past the Iron Age Celts, past the Dark Ages that took hold after the Romans had left, past early kings and queens—and on and on into our own history. Hedges develop in various ways, but some ancient ones are the vestiges of the forests our Bronze Age ancestors fought back as they claimed the land for farming. And they exist now, often with their ditches and banks, forming the barriers between tracts of land, possibly denoting ownerships or boundaries. In the distant past, the boundaries may have been markers between tribal territories; many are now neglected remnants of what they were, but they are still evident.

I could see field maple in the distance, but only hawthorn and oak nearby. In paleoecology (the study of ecology in the past) and archeology you dig down and, as you dig deeper, you go back in time. You count the surviving pollen grains, work them out as a proportion of the vegetation and watch, as the ages passed, how birch trees dominated a landscape and then declined to give way over time to pine, then alder, elm, and lime—how they were progressively removed by people to be replaced by grasses and herbs and, in the uplands, the heathers of moorland. In archeology, you work out the fate of various plants through time, but all those plant communities get conflated through compression of the sediments in your sampling corer, and small differences become compacted together. But here in this field was a single snapshot in time. The difference between this and studies in archeology, or paleoecology, was that changes over long periods of time were irrelevant. The most important thing was what was here when the body was dumped. Analysis in archeology is essentially in three dimensions because time is involved, and time is represented by depth of sediment. Here, the analysis is in two dimensions, length and breadth. I did not have to worry about changes over time at all.

I had another flash of inspiration. Although there was so much oak here, at the field gate why did I get only small amounts of its pollen in the samples? And why had I found so much blackthorn-type, along with black nightshade and dead-nettle? There was none here by the gate. This was not just one hedgerow, but many smaller ones, each stretch distinct from the next, and yet merging with it. It may seem blindingly obvious now, but it was not so clear then. No one really understood how the pattern of pollen fallout varied in this kind of scenario.

I was still gazing along the hedgerow, when the crime scene investigator asked, "Do you want to see where they put the body?"

"Well, actually, I'd like to try and find it myself."

In procession, we walked along the hedgerow but nothing matched the picture in my head yet. I had to keep in mind both the woody and the herbaceous plants that would have been picked up by the offenders. The hedgerow was wrong, wrong, wrong—and then, suddenly, it was right.

"I bet this is the place."

The detectives looked at me. The deputy chief constable's face widened into a smile. "How did you work that out, Pat?"

"Because I have already seen it . . . and all in my mind's eye." I had seen the blackthorn and the field maple, their canopies intertwining, and a hawthorn, with the flowering form of ivy insidiously overtopping its upper branches. Grasses, white dead-nettle, black nightshade, woundwort, docks, goosefoot, and some nettle all contributed to the dense bank between the bare soil of the field and the ditch itself. These had all shed their pollen onto the leaves of the grassy, herb-rich bank. The offenders had trodden in it and carried it back to their car. That hedge and bank had witnessed them setting his body alight and making their getaway. It was informing on them now.

Standing there, with the summer sun beating down and the sloes swelling on the bush, I looked back along the hedgerow. How had I known it was here, exactly here? The fact was: there was simply no other place it could have been. The rest of the hedgerow was . . . wrong. It could not have provided the right assemblage of pollen grains to match the ones I had found in the car. And it struck me then, as perhaps it should have struck me before, that the world is actually so much more heterogeneous, more varied than I could ever have imagined. There was only one place the body might have lain along the edge of this field, only one place where the assemblage I had found in my laboratory, the picture I held in my mind, fitted closely with what I was seeing here at a crime scene. Ten yards along the hedgerow in one direction, the assemblage would have been different; ten yards beyond that, the assemblage would be more different still. There was such specificity in the pollen record.

What surprised me, and now it seems very silly, was that the pollen on the ground matched mostly those plants growing very, very close to where the offenders must have parked their car. The significant exception was cereal. The pollen from that field across the wide road must have traveled at least 400 meters from the crop. That in itself was a revelation to me. There was insignificant influence from plant species that were growing just short distances away. I had little doubt in my mind now that much of the received wisdom in palynological textbooks and papers might need to be revised, at least for some things.

If one stretch of a farmer's hedgerow was so vastly different from another, I wondered what it meant for the assemblages of plants and fungi in woodlands, in meadows, in the grass verges and gardens, along which we walk every day? If one edge of a maize field was so different from another, did that mean that every square meter of landscape was as different from another as one person is from the next? Perhaps the pollen and other microscopic matter taken from a boot, a wheel arch, a doormat, or the pedals in a car were as unique and useful as a fingerprint in determining what someone had touched, what someone had done? Standing here, by a Hertfordshire hedgerow where a man had been left to decompose, I felt my own version of Edmond Locard's "every contact leaves a trace" epiphany: I had thought I had known the natural world reasonably well, but the truth was I had barely scratched the surface. I had been overlooking so much, and the world, which was already so strange and full of wonders, seemed a little stranger and more wonderful still. The information I was able to provide made compelling evidence to indicate that the occupants of that car had contacted the edge of that particular ditch. I did not hear the outcome of the case until very much later but, apparently, my evidence had been an important element in the trial, and the subsequent conviction, of the murderers.

CHAPTER 4

Under the Surface

Not an auspicious start, some might say. I came into the world in the front bedroom of a modest little house in a Welsh mining village, in the depths of a bitter winter, in the middle of the Second World War.

Back then, the world was full of misery—but all in faraway places. We were just half a mile uphill and east of the waters of the Rhymney River, black with coal slurry, gurgling and splashing all the way down to the Severn Estuary. It separated us from Glamorgan, where people were different from us, in our greener land of Gwent. The Black Mountains and the Brecon Beacons were not far to the north, the sea was not far away to the south, and the sparkling waters and verdant valley of the Wye were just a few miles to the east. A short, steep climb up behind the houses would bring you on to wide vistas of bare mountain. Garden flowers were rare where we lived because of the persistent raiding by sheep coming down out of the hills, and there were always escapee ewes trotting, as proud as queens, down the village streets, straggling dirty wool and lambs behind them.

But, forget the winter and the war, the least auspicious thing for me was being born to these parents. My mother was too young, not yet twenty-two, when she had me—although, after three years of marriage, she was anxious to follow custom and have a child. My father was only four years older and, because he worked in the pit, was more useful there than in the army. They were a striking pair. My mother was very small and vivacious, with pale milky skin, light hair, and startlingly blue eyes. My father was a Clark Gable/Errol Flynn lookalike, with jet-black hair, arched eyebrows, a Hollywood black mustache, blue eyes, and a wonderfully rich bass voice. There was not an ounce of fat on him and he had a six-pack that came from hewing coal, not workouts at the gym. He had charisma and women were coquettish in his company. This must have made my mother feel insecure, even though she certainly had her share of admirers. People who still remember her when she was young tell me that she seemed like a film star in that drab little village. She was never seen without lipstick and every hair in place— such a contrast to the young drudges around her who, after the rosy flush of their wedding day, became dowdy as the reality of being a miner's wife hit home. The custom was to hide the hair in a turban and wear a Welsh shawl wrapped around the body to form a sling for the baby, leaving the hands free for work. My mother was determined not to be so shackled—and, as a result, I never had that secure feeling of being close to her breast for any length of time; though I do acutely remember her delicate and distinctive smell. Theirs was a turbulent marriage and, if I felt more generous, I would say my parents could not help it. But I was at the heart of the tumult and it is difficult to forgive the lack of tranquility in your home.

The world then was very different from today, though some of the old ways still linger. Everyone knew what went on in everyone else's lives, or thought they did. My aunties, uncles, and cousins lived in the same street and my father's parents owned the little grocery shop at the top of the hill. My memories are shattered fragments of colored glass, some clear, some vague, others distorted. But my first sharp recollection was sitting, legs outstretched, on the leather backseat of a black car. I distinctly remember being puzzled because my legs were shorter than everyone else's—and I did not like my shoes. When I first told my mother this, she would not believe me until I described the little silky yellow dress with too-tight puffed sleeves and green shoes, each with a yellow stripe from the toe. She was equally stunned when I recounted having my picture taken—being lifted onto a chair in my auntie Eva's front room, wearing a scratchy pink organza dress that was too big for me. It had been sent by relatives from Australia and the family was proud of it. My mother simply could not believe that I recalled it so accurately. "But you were only eighteen months old in that car, and only two when that picture was taken!" she protested. But I do remember; these moments are imprinted upon me. Even now, I look back and marvel that such a young child could be so analytical and so decisive. Never underestimate a toddler—he or she might remember things you would prefer forgotten.

Of course, we are the product of our parents' combined genes, and these influence our innate brain chemistry, but "what we are" also depends on our childhood and life experiences. I think people would say I am outward-going, confident, and assertive, as were both my mother and father, and learning from the behavior of those around me must have played its part. Genes are moderated by memes. Whenever a visitor came to our house, all youngsters were expected to do a recitation, sing a song, or perform one's ex-cruciating rendition of some piano piece. No one was allowed to be shy or refuse. Perhaps it was good training for communicating; teachers were a major export from our little country, and the Welsh seem to have an inclination to the theatrical.

Compared to the young today, our lives were unfettered by worry about how dangerous strangers might be. Any outsider coming into our community was soon identified and, in any case, we rarely saw anyone who was not at least a little familiar. I never saw a black person until I left the Valleys, though some lived in the dock area of Cardiff, and I had never heard an English voice other than my uncle Fred's foreign burr from the Forest of Dean. We had one flasher in the district. Everyone knew him and no one was frightened of him. It was said that he had studied too hard and that it had "turned his brain." Poor man—he used to slouch around in an old army coat and everyone called him "Organ Morgan." I now know why. My best friend and I sometimes caught sight of him when we were on our "adventures." We were avid readers of Enid Blyton's *Famous Five* books and were sure we could, and would, unravel mysteries. We made our own jam sandwiches, pinched cake from the pantry, and set off up the mountain to look for problems to solve. We never found any and would come back with purple mouths, hands, and knees from picking wimberries, the smaller British cousin of the American blueberry, and wet feet from boggy bits of ground. We sometimes got a bit frightened when bands of mountain ponies mugged us for the contents of our little bags. They still roam the hills above the Valleys and, given half a chance, will still accost you for treats.

When I look back on our sheer freedom, I can only feel sad for today's children who are packaged and sealed up, their flights of fancy being satisfied by electronic wizardry. I marvel at how young we were, how far we wandered unsupervised, how nobody felt the need to walk us to and from school—and how entirely normal that kind of free, wild life was compared to today.

I was the sort of child who loved school. That solid, squat, Pennant Sandstone building, surrounded by a big sloping playground and tall iron railings, was everything to me. It was full of gifted teachers who made learning fun, and I was convinced that our headmaster, Mr. Davies, was Jesus. In fact, I *knew* he was. He had a funny nostril and scars in the palms of his hands—and, even though my mother tried to explain that he had been hurt in the war, I did not believe any of it. Mr. Davies behaved like Jesus. He was kind and every one of us loved him. He was the polar opposite of Mr. Probert, who petrified us. He had taught my father in the same school and wore the same starched butterfly turn-down collar and black jacket as he had years before. He had studs in his boots that used to make sparks as they hit the nails in the wooden floorboards when he marched up and down, barking out his lessons; the roughest of boys were subdued in Mr. Probert's presence. Even my father spoke of him with reluctant deference.

I only knew a few boys and girls whose fathers did not work in the pit and therefore, as far as we were concerned, did not have proper jobs. Except for the very few, we were all well fed, well dressed and, on the whole, well behaved; but there was one family where a baby arrived every year, each one seemingly smaller and more delicate than the last. How on earth did they all fit in that little house? They always seemed to be eating bread and jam and, whatever the weather, the boys wore old Wellingtons that left scurf marks around their legs. I remember when they all got ringworm, and had their heads shaved and painted blue with gentian violet, then the only available remedy for such fungal diseases. But we did not make fun, and we were sorry, though scared to get too near. Mind you, neighbors made sure that food found its way into the pantry of that house, and the children were clothed in everyone's castoffs.

All these years later, I still hanker after the way life was in that village. Life was simple; we respected adults and were particularly terrified of "Dai book and pencil." He was the local policeman who got his name from his fondness for pulling out his notebook and pencil as a warning. In fact, most of us were genuinely worried about doing anything wicked. Chapel and peer pressure saw to that. How different it is now. These days, even little ones know about class, color prejudice, sexual deviance, and recreational drugs. These just did not exist for us, and I am glad for my innocent background— I have never been tainted by any of them. Some people were posher than others, but the differences between us all were insignificant, and none of us felt inferior to anyone else. I suppose we were considered to be one of the "better" families because I rarely got dirty, and my father had a motorbike. I had more than my fair share of toys and books and, because of the skills of my maternal grandmother, I had the most exquisitely handmade clothes. My dolls had them too.

Modern genealogy websites have told me that I am descended from the younger sons of farmers who left Pembrokeshire, Radnorshire, and Gloucestershire—leaving the fresh country air to seek their fortunes in the nineteenth-century coal Klondike, and unwittingly contributing to the devastation of our beautiful Welsh valleys. My mother's side was different, though. Her father was

Welsh, but her mother was descended from Scottish farmers and innkeepers' sons who, in the 1830s, with their wives and some children, endured cramped three-month voyages in little ships to New South Wales, where they became tenant farmers and gold prospectors. I come from tough stock—adventurers and hard workers. Even though my Australian grandmother was tiny, she epitomized that strength and robustness—and, of all the people in my life, she influenced me the most.

One of my earliest memories of my father is of him shivering, sweating, and racked with coughs, as this tiny child crept into bed beside him. He had caught pneumonia while lying in wet grass on Caerphilly Mountain doing his Home Guard duty. This was before the National Health Service, and before readily available antibacterial drugs—and I still remember the terrifying reality that he might die. The only thing that seemed to help him at all were lemons steeped in hot water, and I remember the jugs sitting there at the side of his bed. In those days, you either got better or you died. Coughs were suppressed in case you were thought to have tuberculosis—which, because it was then incurable, carried a stigma. My next memory of my father is even stronger: this time, he is looming over me as I lie in bed, his face wet with tears— which I thought very peculiar—because the same pneumonia was in me, his little girl, and I was rapidly fading away.

By then, times had moved on just a little, and the doctor produced M&B tablets, the only relief available. M&B stood for "May & Baker," the pharmaceutical company which, in 1937, was responsible for formulating sulfapyridine. It is one of a range of sulfonamides and, ironically, research into these compounds had its origin in Germany as far back as 1906. M&B powder and tablets became the magic bullets for many bacterial diseases ranging from leprosy to gonorrhea. It also saved Winston Churchill from pneumonia, allowing him to carry on leading the war effort. It is rarely used for humans anymore because of its side effects but, in the 1940s, it saved many from the devastation of septicemia and death that even minor bacterial infections could bring. By the time I reached seven years old, I was already considered to be a "delicate" child and had had my fair share of dosings with M&B.

Another day seared into my mind is a hot and sunny Friday. Fridays meant . . . fish and chips. And as the school bells tolled for the end of morning that day, I was the first up and out of the school gates, scrambling out of breath across the green between the school and our street. Home was only a few hundred yards away—but, when you are seven years old, even this can seem like miles. Still, eager as I was, I hurtled there, knowing that my mother—who hand-cooked the chips for me every Friday lunchtime—would almost be ready.

When I got there, the front door was open, as in all the other houses, and I crept, unheard, into the hall, hiding against the wall between the kitchen and the dining room. What a silly design that house had. The cold larder was off the dining room and not the kitchen. To this day I do not know why, but I waited there, holding my breath, as my mother approached. She hadn't heard me.

I waited and waited until she was almost upon me, and then . . . "Boo!"

My mother reared back in fright, just as I had intended. But the joke was not to last long. She had been carrying the pan of scalding chip fat to the cold larder and I had waylaid her.

As she stumbled back, the chip pan flew out of her hands. The scalding oil arced out of it. I could see it as if it was suspended in the air above me. Then it came down, like a huge wave, across my hair and head, my neck and face. I screamed and screamed and screamed.

I still remember the pain. I still remember the screaming that seemed to be coming from somewhere else but was, in fact, coming out of my own mouth, going on and on and on. I remember a jumble of neighbors, alerted by my shrieks, came running in.

And then I remember nothing until, sometime later, my father burst into the room. To this day, I can see the horror in his eyes as he grasped that the tiny figure, swathed in bandages with only eyes and mouth squinting through, was his daughter. My mother stood back against the door watching, just watching while my father went on his knees to me. I do not think he every truly forgave her for that day. My head would be in bandages for two years. Nowadays, there would have been an emergency ambulance, skin grafts, and the very best treatment but, back then, the NHS was only two years old, and all I had was the local doctor and whatever remedies he could provide. And perhaps that is why the mark of that day has never left me either; even now, seventy years later, I have to arrange my hair carefully to hide the scars.

It was not long after that the repeated illnesses began. Was it coincidence or was it something in my constitution, weakened irreparably by the burns? Whooping cough and measles came all at once. Bronchitis and pneumonia and pleurisy were my constant bedfellows. My lungs never seemed to recover, and soon I had developed bronchiectasis. The disease enlarges parts of the airways, which produce too much mucus. I coughed constantly, was always short of breath, and hawked up blood. My chest ached and ached.

From then on, I was not like all the other children, waking up each morning, wending across the green to school. I stayed home, in my chair by the fire. My teachers visited me, but there was nothing to help disadvantaged children in those days, and my education became loose and haphazard—certainly unsupervised. Thankfully, I was very good at reading. I relied on the books sent to me from the school, as well as my very own set of Arthur Mee's *The Children's Encyclopædia*. What M&B had done for my body, Arthur Mee did for my mind. I loved the stories—Evelyn's discovery of Grinling Gibbons in an East End workshop, Achilles and his vulnerable heel, the discovery of electricity and the properties of amber. By those encyclopedias I taught myself to knit patterned squares, the theory of music, the flags of the world, Aesop's fables, and Roman, Greek, and Norse mythologies. I "invented" slippers from felt, with knitted uppers and elastic to keep them on, and refused to wear anything else on my feet. These books were magical and, on the occasions when I was allowed to go to school, one went with me under my arm.

One person was pivotal to my inevitable recoveries through those months and years racked with illness. She would sit by my bedside and we would read together: *Peter Rabbit, Fuzzypeg Goes to School, Woman & Home*, the newspaper, and my favorite encyclopedias. She took me for walks to see things growing in hedgerows, and look for birds' nests, and taught me what we could and could not eat from the wild. This was my Australian grandmother, who led a strange existence, living with each of her children and their families in turn. We fought for her to come to us, and I loved her more than anyone else on this Earth.

In 1950, still suffering the aftereffects of those burns, a decision was made. My grandmother's cousins, Gwen and Walter, had settled in Rhyl in North Wales after a lifetime in Burma and India, places only just emerging from the old British Empire. My lungs still labored. I still struggled for breath. I still coughed up too much mucus and missed too much school. I had already spent time in a sanatorium and I didn't want to go back there. It had rooms with walls that were pale and shiny, and relentlessly gleaming floors, and there was no escaping the unremitting clean odor of hospital. Food was dispensed by humorless automatons, and the breakfast kippers were so full of bones that the plates were invariably cleared away before I had picked them all out. Every morning and late afternoon, the children were taken to a special room to have their lungs cleared; this involved being bent over a padded beam and being thumped on the back as we did breathing exercises. One of the little girls had a livid scar extending from her shoulder blade around to her front, and I was deeply frightened that I might get one too. I had tantrums when it was suggested that I go back. So, instead, I was sent with my grandmother to Rhyl, where it was hoped that the sea air might bring some life back into my lungs. This was the kind of adventure of which a little girl like me, whose very best friends were her books, could only dream. And to think my grandmother would come with me? The prospect was a joy.

My grandmother, Vera May, was an extraordinary woman. We all remember her as being tiny, thin, and wrinkly. She had white hair and dentures along which she rattled her Mint Imperials. She was formidable, capable, efficient, strict, and overwhelmingly kind. She was my center of gravity, the one I clung to as my parents argued, made up, and fell out around me.

She was born in 1890 on the coast north of Sydney in New South Wales, the descendant of tenant farmers who had pioneered the land in the 1830s. My grandfather, Edmund, a coal miner from old South Wales, had braved the voyage in 1909 when his own lungs had started to deteriorate. He met Vera May just as the Great War was being declared. They fell for each other and married but, as they were looking forward to their first child, they received word that her new mother-in-law was ailing. Left behind in Wales, the old woman feared that the end was nigh   and all she wanted was to have her children around her as she slipped out of this world. She was a powerful, domineering woman and my grandfather felt he had no other choice but to fulfill her wishes. So, in 1916—the year of the Somme Offensive, the year when war was consuming the planet—he put my pregnant grandmother on a boat and to-gether they sailed across the Indian and the Atlantic Oceans to return to his mother's side.

By some strange mercy, my grandparents made it back from the other side of the world unscathed. But, back home in Wales, my great-grandmother did not die—not as she had promised. In fact, she was still clinging to life, a decrepit old lady, when I myself was born. I remember her like a child does one of their earliest dreams, vague and indistinct. But, although she lived on, my grandfather never took Vera May back to Australia as he had promised. They say the call of home is a strong one, and Wales often calls its sons and daughters back. In any case, she gave birth to her first son there, then another. My mother, and a fourth child, followed swiftly afterward.

Then, in 1931, in one of those cruel twists of fate that happen more frequently than they ought, the lung condition, that my grandfather had left Wales to fight, returned with a vengeance. That was the year that he perished—and my grandmother, with four children to depend on her, had to battle on alone. There was nobody else. Nothing to help her. My grandfather left her no pension. The mine owners claimed that the pneumoconiosis that killed him was not their fault, that it was something bubbling inside him all along—and so my grandmother was faced with feeding and housing and caring for four children, without a penny in the world.

So she took to the only things she knew. She stitched and she sewed, for whoever had the need; she took in washing; she kept rabbits in the back garden for meat, and chickens in a coop for their eggs. She allowed cockerel chicks to grow until they were meaty and then killed them herself; and grew potatoes, carrots, leeks, and cabbages in the patches of land around the family house. In the days before fridges and freezers, she would salt and pickle what they did not eat, preserving as much as possible; no one was allowed to waste anything. The whole family had to go picking mountain wimberries and blackberries along the woodland edge when they were in season, and the stained hands and strained backs were rewarded with plenty in the winter.

Yes, my grandmother knew the natural world. She knew which plants were edible and which ones were poisonous; she knew about mushrooms and toadstools; she knew the taste of young hawthorn leaves, the hedgerow plants and berries that made the whole of the world a natural larder. By grit and hard work, and the depth of that knowledge, she made a life in which none of her children went hungry, and she made sure that all of them were given the gift of grammar school so that they could learn the things she could not teach—an amazing feat by any standards. And when, years later, she was needed again—by a little girl with burns, and lungs as damaged as her husband's had once been, she was there, by my side in my bedroom, teaching me the wonderful things that she knew.

My grandmother was an itinerant. After her children left home, and having slaved so hard to keep the roof above their heads, she knew that she could keep the house no longer. But her grown-up children were scattered across the country, finding wives and husbands and forging lives of their own. She traveled to be with them, spending months with one, months with another. She told me quietly and secretly that I was her favorite, perhaps because I was her eldest grandchild—and perhaps she saw, in me, a little something of herself. But I clung to her and the love that poured out of her—and our stay together in Rhyl, which was to last two whole years, was the happiest of times.

In Rhyl, we shared a huge bedroom, my grandmother and I. The house was grand and immaculate and filled with treasures of the East: big, beautiful Buddhas, carved ivory, and oriental carpets. Hyacinths grew in glorious array on the back porch and, because her cousins were old colonials with old colonial manners, there were different dishes for breakfast, tea, and dinner—a peculiarity I loved. Walter and Gwen were nervous of admitting a little girl into their grand house, but became delighted at my good manners and careful handling of anything I was given.

Life with these three elderly people was idyllic for me. We played mah-jongg and I spent many happy hours reading *National Geographic* in the quiet library room.

One day, an animated Vera May whispered for me to come and look at the hedged arch over the path in the rose garden. She lifted me up to see a blackbird's nest with three pale blue eggs nestling in the interior. I shared my grandmother's thrill at seeing this marvel of nature.

When she had gone, I went off to see if the Californian poppies had popped their yellow petals out of their bud sheaths, but I could not get the nest out of my mind. And as the minutes ticked by, a new thought started to occur to me. I do not know where it came from—perhaps only that little, unquantifiable corner of a child that will always ask, *What if?* Whatever it was, a question started hardening. What would happen, I wondered, if I threw my ball at it? And that was exactly what I did. Looking back, I can see it as the futile act of an unfeeling, curious little brat. I watched, with mounting curiosity, as the frantic bird flew off in a chaotic flurry of wings.

The next morning, I found my grandmother standing over bits of nest strewn across the rose garden. The mother bird had devastated her nest. Shards of shell marked the places where she had tossed her eggs out onto the path. My grandmother, this little woman with the white hair, was stiff with controlled rage—and all her anger directed at me. It was not the killing she objected to. My grandmother killed things when she had to. She had been raised the same as all Australian nineteenth-century boys and girls, to know that so many things in the wild can hurt or kill you—scorpions, snakes, and spiders. It amused everyone that she never failed to check behind the lavatory and under the seat wherever she went—because that is a favorite hiding place of the poisonous redback spider. Her fear of spiders was certainly passed on to my mother, and then to me. But her attitude to wildlife was Victorian and, I suppose, uneducated. Anything that looked ugly, or possibly poisonous, could be eliminated. But birds were different—they fell into another category altogether and must not be harmed.

By now I knew what to expect. I hung my head, unable to look my grandmother in the eye. She had learned how to deal with errant children. After all, she had nurtured three boys and a girl on her own. So consumed was she in providing for them, filling her days and nights with stitching and sewing for everyone in the village, with housework, and foraging for her family, that she could ill afford the misbehavior of unruly children—and, later in her own life, my mother would say to me, "You had all the affection from my mother that I should have had," as if the vagaries of her childhood had somehow been my fault. Looking back dispassionately, she was probably right. Yes, although she rarely showed it with me, my grandmother could be as hard-nosed as she was ordinarily joyous and kind.

My punishment for that bird was inescapable, though presented in a very clever way. She made me look her in the eye and admit that I had done wrong and deserved what was coming. "Do you want a slap, go without your treat, or do some work?" The slap was immediate but spoiled my dignity. Going without my Saturday Picture Club at the local cinema was out of the question. So, I never chose either and opted for work. I secretly enjoyed it anyway, even though it was intended to make me bored and miserable. My misdemeanor reported, I was handed over to Uncle Walter who, in turn, passed me the familiar nail scissors. "The grasses are waiting." I followed him out to the back lawn stretching miles and miles to the summer house. It was actually fifty feet long and thirty feet wide but seemed as big as a field to me. Do you remember how the old lawn mowers would scythe through the softer leaves but leave the tough flowering stalks of rye grass still standing?

I had been given this job before and had been surprised and thrilled to find so many tiny, obscure, foreign things between the blades of chopped-off grass. The most productive area for insects, snail shells, animals with too many legs and different tiny colored fragments, which I now know to be minerals, was at the lawn edge. There were plenty of stalky bits to cut, but I could also poke around in the soil with the scissors. I soon realized that the soil was infinitely variable over short distances, and that it was a place rather than just brown stuff. It was home to so many little things with legs. I was made to eat my lunch out on the lawn, which was no punishment either, especially when I found that if I poured my glass of water on the grass and soil, things changed. It changed even more if the drink was hot.

That is what happens, you see, to a child who cannot often go to school, who's sick and sent away and is not up to doing all the ordinary things the other children do. You are left with your own thoughts and allowed to experiment without anyone knowing. And that, I have always thought, is where it comes from—this fascination, not only with nature, but with the stuff that lies under the surface, the aspects of it that none of us can see. Seven decades later, I'm fascinated still.

CHAPTER 5

Conflict and Resolution

Look out of your window. What do you see?

As I write this, I am sitting in the house I have called home for more than half of my life—my detached sanctuary in a leafy part of Surrey. This is where I live and sleep and do most of my work. Were you to look out of this same window, you would see a garden with lawns, separated by rose- and wisteria-laden trellises, stone urns, a bird bath and sundial, and hedges separating vegetable plots from the little orchard and the more decorative beds. The front of the house is approached by a wide drive bordered by flowering shrubs, trees, and the dreaded Spanish bluebells that I have been trying to eradicate for years. You might be forgiven for thinking that this garden has one single "pollen print" by which it can be identified. But you would be wrong. This garden has a multitude of pollen profiles. Look at any garden, any hedgerow, any grass verge or stretch of a country path. To the untrained eye, the greens and browns, whites, blues, and yellows of nature—and all the various shades in between—merge into one. Nature, we often think, is over there, somewhere beyond the place where we are standing. We categorize it all together. But the truth is much more complicated. As I look through my window, I can tell that, palynologically speaking, one corner of the garden is vastly different from any other.

Indeed, in this very plot, keen students I have tutored in the past have proved that a pollen profile from one side of a flower bed can yield a very different picture from that on the other side, just a couple of yards away. A number of samples taken from the base of the ancient hedge, forming the back boundary of our garden, will be utterly different from those taken from the base of the one which separates the vegetable garden from the lawn leading up to the patio. One sample might be dominated by the pollen of hawthorn, field maple, blackthorn, yew, bramble, hedge parsley, and nettles, while another might show an over-preponderance of grass, privet, lilac, honeysuckle, and *Laburnum*; one corner might be full of fern spores, oak, and beech, and yet another be dominated by grass with a smattering of daisy, poppy, cornflower, onion, carrot, bean, and many weeds.

The samples attest to the various kinds of planting, areas of neglect and the garden's diversity. One thing that might strike you as odd is that a set of samples might not be absolutely true to what you see growing in the ground. This is because, unless there are physical barriers, a neighbor's garden, and even a woodland on the edge of the village might also be represented in the mix. Pollen has, after all, evolved to spread. You also need to keep in mind that there will be many insect-pollinated plants that produce so little pollen that it fails to get into the air. These are particularly important to the forensic investigator because the presence of those "rarer" pollen types on your shoes, trouser legs, or pedals of your car would indicate direct contact with them. Think about a foxglove—aptly named because the flowers can fit on the tips of your fingers. An insect has to crawl deep inside the flower to get the pollen and nectar so, if someone has even a few foxglove pollen grains on them, we can say with some confidence that they have been very near to where a foxglove is growing, possibly treading on the soil at its base, or bumping into it in high summer after pollen has been released from the anthers. The pollen of plants like the foxglove or pansy is most likely to be picked up from contact with the plant, or with the soil where their dead flowers drop. If you tread on that soil, the pollen can get enmeshed into your shoe, and it takes laborious effort to remove it.

What is so wonderful is realizing that pollen and spores provide the telltale story of contact—and much more, as I will reveal later. The pollen transferring to your shoes will not be a complete mishmash. All living things live where they perform well, and different plants might all favor the same kind of habitat, though responding to different aspects of it. This is why you might expect to find bluebells in an oak wood, in a coppice of hazel, or even in a hedge of clipped shrubs along a lane; but you wouldn't expect to find them in a dense pine woodland, or growing where the ground is very wet, or on heather moorland.

If you are only moderately observant, you will already possess a great deal of ecological information without realizing it. Where would you expect to find bulrushes? Where would you expect to find docks, nettles, and hogweed, and where would you expect to find honeysuckle or dandelions? They do not occur everywhere, even though many a lawyer has tried to argue in court that this is the case. Plants tend to grow in specific kinds of soil and surroundings, and in communities of others with similar requirements. You cannot grow *Rhododendron* in chalky soil and many other "calcifuges" (plants that hide from calcium) may be growing with it in a recognizable community. *Clematis*, on the other hand, is a "calcicole," or chalk dweller, and will grow with others that either need high calcium or, at least, can tolerate it. *Rhododendron* pollen on a shoe will thus give an immediate idea of the kind of place the footwear has contacted, and certainly give me information about the soil in which it is growing.

It is not only plants that share preferences for specific kinds of habitat. The animals and fungi relying on them might also be found with them. So, if I can identify a specific fungal spore, I might not even need to find the pollen to know that a certain plant was also growing there. A neat example of this secondary trace evidence is that of the primrose (*Primula vulgaris*). The primrose produces very small amounts of pollen—bees have to delve deep inside the flower to get their reward—so we do not often encounter it as trace evidence. But there is a fungus (*Puccinia primulae*) which not only produces masses of spores, but only occurs on the leaves of primrose plant itself. So, if you find even one of the spores you can bet that a primrose was growing nearby. What is more, most people think of fungi being everywhere—mold in the bread bin, rotten apples, and mushrooms on the woodland floor. But the spores of most species do not travel very far at all, sometimes no more than a few centimeters. The frequent moldy objects we see have been colonized by "weed" species, but most fungi are not in the "weed" category. Some only produce spores every few years when conditions are right, and some fungi are decidedly rare. They can be very specific indicators of conditions and places. All of these things help me when I come to visualize the picture of place—the place where a crime has been committed, a body stashed away, or a suspect gone to ground.

Look back out of the window. Now what do you see? If I have done my job correctly, perhaps what you once thought of as small and compact has become something vast and unknown. Your garden, even the tiniest garden, holds so much more information than you can possibly assimilate.

From the vantage point of a tiny pollen grain, even the smallest garden is a vast landscape of different terrains—all interconnected, of course, as ecosystems are; all having a bearing on one another, but all with unique, identifiable characteristics as well. And if a garden is as vast as that, how much bigger is a field? How much wilder and more complex is a hedgerow, a moorland, a woodland? You might think you could get lost, and in a sense you would be right, for when you start looking, there is so much to see. But, as I would soon discover, the complexity and variation is pivotally useful when one square meter of land is different from the next. It becomes possible, by profiling the biological trace evidence, to identify the when and where of a crime, to pinpoint where somebody has been, when they were there, perhaps even—by the traces left on their clothes—what they were doing.

And that was exactly how it was when, some weeks after the Chinese Triad case in Hertfordshire, I was in my laboratory when the telephone rang again, and that same Glaswegian gruffness that now had a name—Bill Bryden—greeted me again. "Pat," he said, "if you haven't had enough of us, something's come up . . ."

It is not always a murder. Murders make headlines, but every day somebody ruins someone else's life somewhere by selfish or evil behavior. And, on an overcast day in July that year, I found myself in a neat little square of shops, with flats above, in Welwyn Garden City, staring at the token municipal flower bed set into the paving stones. The roses planted as ground cover were pink and fragrant, but with dense, sharp thorns to keep off unwelcome residents. It struck me as being quite sweet and strange that young lime trees had been planted in the same plots as the roses. They make strange bedfellows. In nature, the chance of finding much lime pollen in the same sample as a lot of rose must be utterly remote. Generally, the pollen of the rose family is never abundant. It would be easy to conclude that this was an artificial habitat—a garden or park, although shopping center would not be the first to come to mind.

A girl claimed that a boy had threatened to kill her unless she agreed to have intercourse. They had been to a youth club together but, as in so many cases, his youthful testosterone surges had been uncontrollable. Most would be inclined to believe her as she was ravaged by deep scratches and a large body-shaped impression had been made in the rose bed. No one in their right mind would lie in that bed of thorns willingly, but the young man steadfastly denied her claims, and I was asked to test his story.

My "eureka" moment with the hedgerow in the previous case might have opened my eyes to new avenues in palynology, but I had since returned to my laboratory and day-to-day work without any real sense of expectation. Academic work can be like that: a strange diversion, a weird digression, and then back to fieldwork, departmental seminars, the daily grind of counting pollen and spores that may have fallen into a pit or ditch in Roman times, or deciphering what the collection of pollen grains in a bronze vessel, dug out of a fenland grave, might mean.

Yet, in the weeks that followed, my mind wandered back to the murder. International gangs and money laundering, and a hapless attempt to hide a dead body: so far, so colorful—but the case into whose depths I was now staring was not nearly so absurd. This was just another of the ordinary evils people perpetrate against each other every day. I forced my thoughts back to this alleged rape and tried to imagine the scenario playing out in this little square of shops and flats—such an open and overlooked place. The boy and the girl had spent the evening together at a local youth club. After dark, walking home together, the couple had lingered here while their other companions departed, laughing, into the night. She said they had held hands and kissed. Perhaps it had gone further still. But when the boy pressured her for sex, she said no, and here their stories diverged.

Two different versions of events: in one, the boy and the girl go their separate ways, each off to their own homes; in the other, the boy pushes the girl backward, forces her down among those spiteful roses on the square, pulls off some of her clothes, and presses himself upon her.

What was the truth?

Rapes like this are extraordinarily difficult to prove. Where DNA evidence is not conclusive or does not appear to have been left at the scene, where witnesses cannot corroborate, or video footage does not allow, all the police have to go on are two stories, each battling with the other. Frightened or liar? Vindictive . . . or victim? Providing solid evidence for one or another can be almost impossible. And that was where I, on my second tentative "case" for the police, was to step in.

I had learned some (as far as I am concerned) dramatic facts in my first case: there is such specificity in an assemblage of pollen grains that one can envisage the place from which it was collected. Of course, this is something that I had been doing in archeology for years, but I had never had any way of corroborating my ideas. Palynologists working on ancient materials can never really be sure that what they see in their mind's eye is accurate. The past is past and gone. Now I was excited to realize that I could check the validity of my interpretations—and I had discovered that the landscape can be re-created so exactly and, on such a small scale, that the potential it has for forensic investigation was vast. It was too exciting for words. In Hertfordshire, a group of organized criminals had been convicted by the pollen they had brought back on their feet to the car. Here, as I looked into the flower bed, my mission seemed clear. The boy admitted that he had been with the girl in the square, but denied that he had tried to rape her and declared that he had had no contact with any flower bed. Perhaps his shoes and clothes would tell another story.

Both rose and lime rely on insects to carry out pollination and both plants produce relatively small amounts of pollen. This little fact was going to be significant in whatever I did. I knew that if rose pollen turned up on clothing, it was a good bet that the garment had touched a rose plant, or the soil immediately beneath it. If the plants had been widespread and wind-pollinated, it would mean that they shed their pollen farther afield and it might have been spread across the square. This would have made it more difficult to declare, with any confidence, whether the boy had been in the flower bed itself, or just hanging out innocently nearby. Or would it? I needed analysis to find out.

The hypothesis in cases like these might seem simple: if the pollen derived from the flower bed matched his clothes, in content and proportion, it would be likely that he had contacted the flower bed. But in reality, the challenge is much more nuanced, and getting a black-and-white answer would depend on the nature of the pollen itself.

The morphology of rose pollen is so similar to its close relatives that, if any were about, it would have been difficult to separate it, with any confidence, from bramble, apple, or hawthorn. Finds of it in ancient deposits are comparatively rare when compared to other plants so, if we found rose pollen here, and on the boy's clothing, we might suppose that they could have come from the same place. Lime, too, is not as commonly found as that of, say, oak, hazel, or pine. As far as I could see, there were no other lime trees anywhere in the vicinity of this part of the town so, if I found lime pollen on the boy's clothing, there was a good chance that this was the place of contact. Because of the small amounts of pollen produced by the rose, and the fact that it is rarely found in any palynological profiles, we can consider it as being palynologically rare. The plant itself may be quite common but its pollen can rarely be found because it is collected and carried around by insects. Clover, with its heads of tiny flowers, falls into the same category. This is the way they have evolved, and forensic palynology can exploit this convenient characteristic.

This was only my second case and I was on a steep learning curve. With the police, I made a grid on the flower bed and systematically collected rose leaves and soil from each square. By doing this, I would be able to determine just how well each species was represented in the putative crime scene. Each sample was numbered, dated, timed, and placed in paper bags which were printed with police references on the outside. Detailed notes were made by the crime scene investigator while I drew a diagram of the flower bed and made my own notes about the sampling. I had been told that every little bit of scribbled paper must be kept if a case goes to trial, so I am always careful to save everything that might be relevant.

There were no physical barriers, such as walls or other shrubs, so the process of sampling the square was straightforward—and, by the end of it, I had the necessary "comparator" samples. These would enable me to test their similarity with that of the boy's clothing. Invariably, a suspect will claim to have picked up pollen and spores from other places, the so-called "alibi" sites. If this happens, it is essential to visit those places and collect "alibi" samples for comparison with the crime scene samples, and from exhibits seized by the police. The only exhibit available to me was the suspect's bomber jacket and his shoes. There was never any doubt about the identity of their owner; on this, at least, both boy and girl agreed. Getting the pollen out of the soil from the square was going to be relatively straightforward—I had done it time and again. In my archeological work I had extracted good material for analysis from pottery and bronze artifacts. But I now had to think about doing the same from fresh leaves, a pair of shoes, and a bomber jacket, which was a mass of synthetic fabric and plastic. I would have to be inventive. I had certainly been used to that in both ecological and archeological work.

First of all, I had to envisage what went on during a rape. From what the girl had said, the boy had forced her down and lain on top of her. This meant that his elbows and the front of his jacket might be the best place to look for evidence. He would have knelt too, so his knees would have had direct contact with the leaves and soil. I surmised that the posture of the suspect during the putative act would have involved his knees, chest, elbows, and toes. His chest, elbows, and toes could be checked against his jacket and shoes, so there was a good chance that we would be able to accept or deny the suspect's story.

My logic was that if the defense claimed the pollen from the jacket had come from the air and by casual contact, the shoulders, back, and front should yield very similar profiles. In other words, samples from the back of the jacket would provide good controls in testing the hypothesis that the observed profile from the front had been picked up from the flower bed. If the front resembled the flower bed and had minimal similarity to the back, this might be considered good evidence of contact. I tried not to predict the outcome; I just wanted to see the results. Back in my lab, making sure not to allow the various parts of the jacket to contact each other, I cut the sleeves away, and then the back. I was then left with two front pieces which I could combine into one sample.

In those early days when I was taking tentative steps into the world of forensic detection, I knew very little about forensic protocols—those rules and regulations designed to make sure that no one had tampered with evidence, or that evidence was corrupted, or otherwise made unsafe during police handling. I had never even been in a forensic laboratory but, of course, I was well trained in laboratory procedure and, because of my background in microbiology, I certainly understood the basic principles of septic technique and prevention of contamination. It was just common sense to me, and I did not need any special tutoring in that.

The brutal truth is that, even today, some police officers are woefully underprepared when it comes to handling some classes of evidence. They are taught how to take DNA samples and to prevent cross-contamination. They remember Locard's idea—"every contact leaves a trace"—and know that contact between two pieces of evidence can render that evidence completely worthless in court. But, even now, after all these years of teaching them, they still cannot seem to get their heads around the constraints and requirements of environmental sampling.

Many people have the wrong concept of forensic science, and the term is used in a very sloppy and inaccurate way. Perhaps you think "forensic" simply means "carried out with great care," but the true meaning of the word is actually much more specific. Court cases in ancient Rome were held in the Forum, and the word "forensic" is derived from the Latin *forensis*, which means pertaining to the open court or public. So when we speak of "forensic" what we actually mean is that any evidence produced will be pertinent to a court case. If a piece of work is not carried out with a view to it having some bearing on a court case, it is not forensic at all. And this is why such careful handling of the boy's jacket was an absolute necessity. Whatever I discovered had to be uncorrupted and uncontaminated, and my training and experience in laboratory work over so many years meant that this concept was deeply ingrained. I use the same principles in my kitchen at home, and I am sure that no one would get food poisoning from my preparation of food.

I did not need to evoke the picture of place in this investigation. I already knew what the putative crime scene looked like. But visualizing what had happened there in the square was critical to deciding how best to approach this, even though it did not need much imagination. I had already worked out which parts of the jacket and shoes needed to be processed and investigated. But, how was I going to get the palynomorphs out of the fabric of the jacket? I had never done this before. Years later, at a conference, commercial representatives were keen for delegates to examine their wares, and one being demonstrated was a microscope with a special attachment emerging like a long antenna that could be poked into restricted places. Its lens was powerful enough to see pollen grains directly without mounting them onto slides.

I quickly went to the outside of the conference center to get some pollen in the garden, and came back with some tulip anthers that were full and ripe. The representative dabbed some pollen onto a little piece of cloth and directed the antenna with its strong lens toward it. An audience had gathered to see what would happen and to everyone's amazement, the pollen grains seemed to come alive. They wiggled and danced down into the weave of the fabric. I was astonished, but it made me realize why it was always so difficult to get palynological trace evidence from woven material. Mind you, this difficulty also means that pollen and spores provide excellent trace evidence because, unlike fibers and mineral particles, they become deeply embedded and are not easily lost. My subsequent work has shown that it can remain in all sorts of fabric for many years, and this is why palynology is such an excellent discipline in "cold case" evaluations, where analysis may be needed long after a crime has been committed.

Only after searching through the literature did I finally understand the reason for the pollen's dancing behavior. It is all down to electricity. Pollen is negatively charged and is, therefore, attracted to any positive charge. Bees are positively charged and (this is wonderful) they are attracted to flowers with the strongest negative charges. I had always assumed that the reason pollen stuck to bees was because of their hairiness and the stickiness of the pollen. Well, that may be part of the story, but there is little doubt that static electricity is also important in the transfer from flower to pollinator. The negatively charged pollen jumps onto the positively charged bee through electrical attraction.

Since those early encounters with the unknown, casework and experiments carried out by my M.Sc. students have shown me that many objects, fabrics, and substances will attract pollen strongly: human hair, fur, feathers, nylon and other synthetic fibers, wool, fleece (which is made from recycled plastic), plastic itself, and so on. Seemingly clean exhibits can have a heavy pollen load which, of course, is utterly invisible—but I am acutely aware of it, and nowadays I will never neglect any object in a criminal case. I once retrieved a few spores from a flashlight which had been used by a murderer while burying his victim. That flashlight yielded just a few pollen grains and spores but this was enough to tell us that he had laid it down at the edge of a fallow field—not much more, but that was enough for the clever investigator who eventually arrested him. Palynological intelligence can be powerful.

All of that, though, was in the future. Right now, and without knowing any of this, I had to devise a way of getting evidence out of that jacket. I did not know where to start, but at least I was armed with a solid grounding in the types of things I was expecting to find. Pollen grains and plant spores are remarkably robust. They each have an outer wall made of a complex polymer called "sporopollenin," and we are still not sure of its exact chemistry. As some paleobotanists and geologists will attest, in the right conditions it can last for millions of years. A friend of mine, Professor Margaret Collinson, retrieved a whole bee out of Cretaceous deposits, and the pollen sacs on its legs, with all the pollen grains beautifully preserved, were clearly visible. The pollen was about 100 million years old but had been preserved in sediment that had consolidated to form rock. It is not only amber that can preserve insects.

The robustness of those pollen grains had to be an advantage in my attempts to get them out of the fabric. One answer was to try to dissolve the garment in strong acids, leaving the more robust pollen grains and plant spores behind. This might have worked on cotton, linen, or any other natural plant fiber—even if it was reconstituted, such as rayon or viscose—but it would not have been effective on synthetic fabrics such as acrylic, nylon, or polyester, which are manufactured from the by-products of the oil and coal industry. The jacket was made of some fleecy material, effectively from recycled plastic bottles, and I was pretty sure that I could not dissolve it away. In any case, it would have been a horrendous job and, with that amount of fabric, virtually impossible, as well as prohibitively expensive. No, there had to be another answer.

When I found it, it had been staring me in the face. I had spent much of my working life in laboratories, but I did not find my answer there. I might have been a scientist, but I was a housewife as well, and once upon a time I had been a mother. I certainly knew how to get dirt out of clothes. One needs a surfactant such as detergent, which lowers the surface tension of the water so that it can penetrate and lift and flush away embedded dirt and particulates. So what is the difference here? It seemed so incredibly obvious. This was exactly what happened in my washing machine every time I used it.

Perhaps you have heard of the principle of Occam's razor. Well, this was a perfect example of it coming to life. William of Ockham (1287–1347) was a Franciscan scholar who favored "the law of parsimony." He said that, when we solve problems, the simplest solution is often the right one. Aristotle had had the same idea, many centuries before, and it is certainly a useful one in science, especially when grappling with complex scenarios and multiple possibilities.

Detergent seemed the simplest solution, so detergent is what I used. I was worried that, when wetted, microbial activity might be stimulated in the fabric and pollen might start to decompose, so I needed some kind of disinfectant. Again, I was worried that the ones we use domestically might affect the palynomorphs. I needed something gentle, something that would not oxidize the palynomorphs; how about medicated shampoo? First of all, I needed to check that medicated shampoo did not contain pollen grains. It was highly unlikely but I had never looked, and I doubt if anyone else had done so either. I found it was sterile of anything, and so I had a good surfactant and disinfectant in one. As I was to later discover when dealing with samples from decomposing corpses, the use of medicated shampoo had an extra bonus. It sterilized samples contaminated with bacteria and, in the process, removed the sickening smells as well. What a wonderful product for the forensic palynologist.

I bought some new stainless steel bowls and sterilized them with neat bleach. This oxidizes away many organic molecules, as well as bacteria and fungi—and, by the way, the tannin coating in your teapot. I then used the minimum amount of very hot deionized water which, although supposedly sterile, would need to be tested for pollen contamination. With dilute shampoo, I worked away like a washerwoman. I agitated, rubbed, and rinsed each part of the garment, finally flushing it through with deionized water. This was certainly not rocket science, but I could not think of a better method.

Each sample resulted in a gray, murky suspension. It was

interesting to see how dirty this seemingly clean jacket had been in reality. By the end of my work, I had five samples—two fronts, two sleeves, and the back of the jacket—to compare against the ten samples I had taken from the flower bed, five of foliage and five of soil. I washed the leaves just as I had washed the fabric of the jacket. After processing, I made the microscope preparations and could not resist just quickly scanning each one to see what they were like. My extraction by washing had certainly worked well, and the preparations were rich in palynomorphs. Within a few minutes of these cursory scans, I knew the answer to the case. But scanning results would not be adequate for presentation to the court, and the tedium of identifying and detailed counting everything I saw was essential.

When starting analysis of a slide, I always start my first transect at the left-hand top of the slide, slowly passing down to the left-hand bottom, viewing continuously under the microscope, stopping only to adjust to a higher magnification for more definite identification. This needs immersion of the lens in oil, and the use of phase contrast settings on the microscope. If any pollen grain were not easily identified, its coordinates on the slide were noted so that later I could scrutinize it more closely with the aid of my extensive collection of pollen reference slides. So, from the top edge, down to the bottom, moving over to the next field of view, then back to the top to start the examination of the next transect, repeating this over and over, covering as much of the slide as possible to remove sampling bias.

In archeology, as in forensics, this is a painstaking business. I have lost hours, days, weeks of my life to the microscope. The concentration it demands is immense; everything seen must be considered and cataloged, whether it is a pollen grain, fungal spore, fossil spore, or other microscopic organism. Focusing so intently on such tiny things is its own form of exquisite torture. Sometimes it can take an age before you start to build up a mental picture of a place. You are looking for patterns, concentrations of one plant's pollen or concentrations of the next. But today a quick scan of the slides, before any real counting could begin, suggested exactly the assemblage I was looking for: rose pollen, each one with three deep furrows curving from the poles to the equator, where bulging pores were situated, and lime pollen, so utterly easy to identify. It is flattened pole to pole and has three inverted pores around its equator, the outer wall decorated with tiny craters. This grain is a favorite of students just starting out in palynology because it is so easy to recognize.

Another palynologist might have looked at the amount of lime and rose pollen on my slides and deduced that it was not enough to make any conclusions—but this only emphasized, to me, how little rose and lime shed into what we call the "pollen rain," the pollen and spores that fall from the air.

The proportion of rose pollen in the flower bed was 10 percent, and on the jacket front was 7 percent, for lime the result was 18 percent to 15 percent. The percentages were close enough to convince me. The likelihood of picking up that amount of these taxa without direct contact was infinitesimally small. The very small amounts of rose and lime pollen retrieved from the flower bed simply demonstrated how little pollen these plants released from their flowers. The rest of the pollen profile on the garment and in the flower bed was quite diverse, but interestingly, the same assemblage of taxa was present in each.

It was intriguing that there was no rose or lime pollen on the toes of the boy's shoes, nor indeed much of anything else. Then I thought back to that place. The flower bed was so small that his feet had remained on the paving slabs surrounding the bed, and he would not have picked up much from those slabs.

The back of the jacket yielded very few pollen grains, and no rose or lime were found. It seemed clear to me, then, that the back of the jacket had not contacted the flower bed as the front had done. It proved to be a good control because it demonstrated clearly that any pollen coming from the air, or other sources, was very sparse and, indeed, did not match the flower bed. In my view, there was a high likelihood that the jacket front and the elbows had been in contact with the foliage and soil in the flower bed.

The boy was lying—the girl had certainly not scratched her own skin in order to claim rape falsely. False claims of rape do happen occasionally, and I have saved one or two young men from jail with similar kinds of analysis to the first of my cases.

Building up a reference collection in my early palynology days was a happy time. It involved gathering flowers in the field, accurate identification, and visits to herbaria and museums if they were generous enough to give me a few anthers from specimens. My reference collection is so very precious, and I would feel insecure without it. Indeed, one never stops collecting and comparing. The process of identifying and counting each pollen type in each sample is mind-blowingly tedious, although moments of excitement come when unusual pollen or spores are found. The ones I hate most are the small, oval-shaped ones, with three furrows and a fine network across the surface of the grain. They are the pollen equivalents, or almost, of difficult groups of plants like blackberry or dandelion, where tiny groups of specialists are the only ones who can name them confidently—or the LBJs (little brown jobs, as my husband calls them), small brown fungi with few distinguishing features.

It is also difficult knowing when one has counted sufficient pollen to make a case. Sometimes, one can produce evidence from relatively few, and other times one might have to count thousands. A textbook needs to be written about these problems and, perhaps one day—who knows.

After the enforced but, in reality, unnecessary counting of samples from this boy's clothes, I calculated the relative frequencies of various pollen and spore taxa, drew up some bar charts to help the police understand what I was getting at, and they passed them to the Crown Prosecution Service and the boy's defense attorney.

His attorney must have sat him down with his parents and lawyer and showed him my diagrams and tables of figures, with accompanying explanations. He must have been stunned that his jacket had revealed the truth about his attack, and he reluctantly confessed. The girl was spared the agony of being put in the witness box in a court of law and having to relive that night under the scrutinizing eyes of the jury, the public, and the press. The testimony of the thorny rose and lime trees in the square saved her that ordeal.

What I had done was so simple—soil and dirt, a dash of medicated shampoo, a few inventive ideas, and a healthy dose of common sense backed, of course, with years of study, hard-gained knowledge, and experience. My first two cases had provided valuable information that had made a difference. Perhaps forensic palynology had a future.

CHAPTER 6

"I put it to you
that you were there."

I was not the first person to realize that botany could be useful in the world of forensics. Identification of the wood used to construct a homemade ladder was instrumental in securing the conviction of Richard Hauptmann, who kidnapped and murdered the baby son of the famous aviator Charles Lindbergh in 1932. Hauptmann's trial was one of the first media circuses—one of the first "trials of the century"—and his conviction rested upon the work of Arthur Koehler, a wood anatomist from Wisconsin. By identifying the genus of tree, the milling pattern, and the direction of growth, he was able to prove that the timber used in the ladder the kidnapper used to steal into the baby's room was taken from the attic at Hauptmann's own house. Hauptmann was strapped into an electric chair in April 1936, paying with his life for his crime, and all down to the testimony left behind by a piece of wood.

Nor was I the first person to identify pollen grains in the quest to solve a case of murder or missing people. The first record we have of a police investigation using palynomorphs goes all the way back to Austria in 1959, and the disappearance of a man sailing down the River Danube. Without a body, police had little to work with until the investigation turned to Wilhelm Klaus, a respected palynologist from the University of Vienna. Klaus was given the boots of the missing man's close friend and, by microscopic analysis, was able to identify an assemblage of modern, well-preserved pollen of spruce, willow, and alder trees. But there were also fossilized grains of hickory pollen. The distribution of these deposits was specific; they were peculiar to a small area located twelve miles north of Vienna. Klaus told the police where to look and the suspect was so shocked at the evidence that he confessed and led investigators to the body. Klaus had envisaged that place using his knowledge of botany and geology of the region.

These people, among others, went before me with their forays into forensic ecology but, in Britain, the potential of botany to contribute to criminal investigations was hardly exploited before the Hertfordshire case, and most countries of the world are still unaware of what has been achieved. My challenge, across the years, has been to bring a range of specialisms together to form the basic tenets of forensic ecology, and to share this knowledge as comprehensively as I can.

Palynology was already a well-established discipline by the time I made these first tentative steps, but transporting an academic discipline into the world of police work presented a succession of unique challenges. The truth is, those challenges have not stopped to this day, and they never will. The range of variables in any environment is so wide that each new murder, each new missing person, vicious assault, or rape presents a unique set of circumstances. A discipline like ours advances by degrees—it is cumulative. The natural world is a complex set of different interacting systems, and to perform at their best, a forensic ecologist must have good ecological training and understand the interaction between organisms and their environment, both physical and biological. They usually have one or two areas of considerable expertise, for example botany, palynology, and soil science, but also some knowledge, and certainly an appreciation of, entomology, bacteriology, mycology, parasitology, and zoology as well as chemistry and statistics.

By trial and error, I have spent the last quarter of a century developing the protocols by which the discipline is now defined, but the simple truth is that no two situations are the same, and there are very few fixed rules and protocols that can apply to them all. Often, I have been "flying by the seat of my pants," having to invent ways of retrieving palynomorphs from various objects and materials while at crime scenes, or in the mortuary. Eventually, from sheer, grinding experience, I was able to publish a list of protocols for forensic palynology. These contain nothing fanciful, but my findings have, on many occasions, demonstrated that accepted wisdoms in classic palynology have had to be jettisoned to cope with reality.

The plant kingdom is vaster than most of us can imagine. Excluding algae and mosses and their allies, it is estimated that there are about 400,000 species of plant, with about 370,000 producing flowers and pollen. The rest produce spores. New species are being discovered regularly and, in 2015, over 2,000 new species were identified. When it comes to fungi, we are in a different ball game altogether, and estimates of species are in the millions. Every year, the number that is new to science is huge, and it seems that the count is limited only by the availability of competent mycologists—those who study fungi. Even in my own forensic cases we have identified several new species.

We will never know how many animals, plants, fungi, and other organisms there are on this planet. And it is chastening to think that most that ever evolved and lived have now become extinct. The vastness of the biological world today is a small remnant of what Earth supported in the past, and no single lifetime would be enough to acquire the skill needed for identifying even a small part. But, being able to accurately identify organisms, or parts of them, is important to a good biologist, and absolutely essential for forensic work. Accurate identification of families, genera, and species can represent the difference between life imprisonment or freedom for someone.

The specks and fragments of the environment that get picked up when you contact it, whether it is on your clothing, footwear, hair, garden spade, or vehicle, are proxy indicators of that environment. They are the trace evidence that may link you to a certain place at a certain time. It could be a whole stand of vegetation; it might be an entire plant, or part of a plant; it might be roots, wood, bark, twigs, leaves, stem flowers, or fruit; more likely, it is smaller still—pollen from conifers or flowering plants, or spores of mosses, ferns, and fungi. The tiny, invisible proxy indicators like pollen and spores are particularly valuable because they are invisible to the eye, so you cannot see them to remove them. You are unaware of their presence, and it would be hard to get rid of them even if you were. They can be a clandestine record of where you have been, and what you have been doing.

Fungi are particularly interesting because they can act as secondary proxy indicators. They may be growing on a plant; you may not pick up the pollen of that plant, but you might find the spores of its fungal parasite, which may always be associated with it, as in the case of the primrose mentioned earlier. In principle, by finding an assemblage of palynomorphs on a suspect or victim's clothing or, in other cases, the car in which they traveled, the tools they used, or even the insides of their bodies themselves—we can conjure up an image of the kind of landscapes through which they have traveled or what happened to them. If this sounds simple at first, a case of joining the dots and marrying one imagined landscape to another, the truth is anything but. In reality, the detritus we salvage from a person's clothing is a chaos of different microscopic particles, many of which will be unidentifiable and unhelpful in terms of constructing our landscape.

The outside of the car in my first case in Hertfordshire showed the many and varied landscapes through which the car had traveled in the past months, and the same can be true of shoes, jackets, coats, and jeans which have not been regularly changed or laundered. Finding and imagining the right landscape, filtering the useful data from those that might lead into blind alleys or dead ends, involves nuances it has taken me decades to understand. It demands knowing not only one set of palynomorphs from another—when the differences can already be infinitesimally small—but also the ways in which palynomorph assemblages can lure you into thinking the wrong thing. It is vital to know the way pollen and spores are dispersed from their parent plants and fungi, flowering times, the kind of soils and conditions they proliferate in best, and the kinds of plants and fungi that tend to grow well in concert with one another, benefiting from the same kind of conditions.

Plants that thrive in similar conditions will be found in the same habitats. They can form recognizable communities and from them we can extrapolate a great deal. If somebody gave me an assemblage of palynomorphs like bulrushes, reeds, sedges, gypsywort, and purple loosestrife, I might recognize a pond, lake edge, or ditch. The species of bulrush might even give information on the degree of flow in the water. Oak, hazel, ash, bluebell, and wood anemone might evoke images of our wonderful bluebell woods which are so peculiarly characteristic of the British landscape. The point is, plants do not grow just anywhere. Everyone knows that you cannot expect banana trees to grow in the wild in Norway, any more than you would expect to find cacti in the arctic circle, or polar bears in the jungle. These have their specific requirements and they would not thrive out of them naturally. And yet this is one of the common mistakes made by lawyers I have come up against in court. They know little of plant science or ecology, and I have been challenged with "dandelions are found everywhere, aren't they?" Well, of course they are not and, in fact, finding their pollen gives us quite specific information.

Knowing the peculiarities of species allows me to understand the assemblages of pollen grains retrieved from a soil sample in terms of habitats and ecosystems. It allows me to see the kind of place represented by the pollen and spores, and the greater the number of different taxa in the assemblage, the sharper will be the image. These are not simple processes; there is no step-by-step procedure to follow to lead you from a chaos of raw data to an absolute answer. With so many aspects to hold in mind, the best forensic ecologist must perform a kind of informed intuition.

As if to prove the point, in 2009 I arrived at the thirtieth anniversary conference of the Association for Environmental Archeology in York. Here, environmental archeologists from across the world gathered to do what scientists do at conferences like this—to pool expertise. By now it had been some years since I last worked in archeology, but I had resolved, nevertheless, to deliver a paper and, taking the floor, I presented a spectrum of pollen taxa I had obtained from a forensic case. The results were from a case commissioned by the RSPCA against badger diggers. These men, who can only be considered to be cruel thugs, put their dogs down into setts to find badgers. Both animals invariably inflict horrendous injuries on one another; in ordinary circumstances, the dog would come off worse—except that, once located, the men simply dig out the hapless badgers and kill them. This is great sport, isn't it? It is, of course, against the law in the UK, and in this particular case the RSPCA was hoping to get a conviction to deter others.

I had been given soil-laden spades and samples from the top and center of a badger sett, and then had to compare the palynomorph spectrum of the spades with those from the badgers' home. Analysis showed that they matched well, but the cherry on the cake was finding a rare spore in both the crime scene soil and in the soil on the spades. This particular spore, so rare that neither I, nor Judy Webb, my colleague who is absolutely brilliant at identifying pollen, flies, and many other things, had seen it before. It was a turning point in both this case and, in fact, the course of my life. Not only did it become a key piece of evidence for the prosecution; it was how I got to know my new husband. I had not met him up to that point but I was lucky that, while I was in the throes of this case, I literally bumped into this lovely man at a memorial service for one of my dear botany teachers from King's College, the renowned Francis Rose. The service was held at Wakehurst Place, the country establishment belonging to the Royal Botanic Gardens, Kew and, after the inevitable ceremonial tree-planting, this very well-dressed man and I were walking together toward tea and cake through the dappled light of the woodland when I spied through the trees what I thought was *Russula emetica*, a lovely red-capped fungus.

"Oh yes," said my companion, "it's *Russula* but not *emetica*, I think."

"Oh, you know about fungi?" I asked.

"Yes, a little," he said modestly.

That was of immediate interest to me as I desperately needed someone who could identify difficult fungal spores.

"What is your name?" I asked casually.

"David Hawksworth."

It slowly sunk in. "Not *the* David Hawksworth?"

I was shocked because in my mind this man should have been about ninety. The work of this world-renowned mycologist had been in literature for years, and was certainly on my list of references for my students at King's College, and at least two of his books were on my bookshelf. Yet here was a very interesting man of about my age, with twinkly eyes and a boyish grin. When I got back, I told Judy Webb about my chance meeting and she was excited. "Oh, Pat," she said, "you must cultivate him. He'll be very useful to us." The rest is history and I married this clever, knowledgeable man three years later.

It was through David that we discovered that the weird spore from the badgers' sett was from a fairly rare truffle which grows on the roots of oak trees. Like dogs and pigs, we learned from this case that badgers obviously love truffles. One could only suppose that they had foraged in the roots of the oak tree about 100 meters away and brought truffles back to the sett. There were the spores as testament and, my word, they endowed a rare and convincing component to the palynological profile. They were not found in any of the other samples, and the truffle gave the crime scene profile a distinctive identity. When the spores were also found on the spades, they provided evidence with much greater potency than the pollen alone had given.

At the archeological conference, I presented the attendees with an assemblage of pollen from this case, dominated by oak, with some elm, maple, lime, hawthorn, ivy, and honeysuckle—along with some buttercups, foxgloves, grasses, docks, and ferns. I then asked them what kind of place they thought this was. A few hands were tentatively raised—no one wanted to look silly—but all those who called out said roughly the same thing. "It's woodland," they said, "from the edge of a glade."

This was a perfectly reasonable assumption—but it was also quite wrong. This was not woodland at all; these samples had been taken from open pasture, about 10 meters away from a very old hedge and a hundred meters from an oak tree. The landscape as a whole had been open, but divided into neat fields, all bounded by mixed hedges. I had confirmed my long-held suspicion that many palynologists still know little about how pollen actually disperses across the ground in most habitats in spite of the heavy weight of literature about the phenomenon.

In the United Kingdom, historically, the countryside has been developed by different kinds of management and two distinct types are recognized. Counties like Essex, Sussex and Suffolk have wood pasture (bocage) country. This is the classic, pretty, mixed arable and pastoral countryside of England, with small fields enclosed in species-rich hedges, with some standard trees. The other kind is fielden (champagne) country which, in the historic meaning, simply means a landscape of wide, open fields, where arable farming involved division of the land into strips, so typical of Lincolnshire, Leicestershire, Wiltshire, and many others.

All those daring to answer in that conference room had got it wrong. They had assumed woodland but, in actual fact, it had represented bocage country, with the "woodland" pollen and spores coming from the hedgerows enclosing the fields. The herbs were survivors of grazing, or those growing in the shelter of the hedge boundary. This, of course, was the lesson I had learned in my first case with the police—and it is the sort of thing that constantly complicates assumptions. The palynologists in the room were absolutely gobsmacked that their interpretation was so far off. It also demonstrated the degree of cognitive bias that, through default, prevails in much thinking in environmental interpretation.

Every last one of the palynologists in that room had jumped to the kind of wrong conclusion that, in my own field of work, might have serious consequences for a victim of crime, or somebody accused. In forensics, where people's lives and freedoms matter, the stakes are very different. And that is why re-creating possible landscapes from the palynomorphs deposited on clothes and shoes is not as straightforward as just tallying up the pollen taxa, spores, and other particulates left behind. One cannot just read the existing palynological literature and assume that a certain assemblage will represent a specific landscape. One cannot act like an automaton, using other people's published interpretations each time; it would make pollen and spore interpretation equivalent to "painting by numbers." Only by knowing about the natural world we live in, both on vast and microscopic scales, can we be confident of getting close to the truth.

There are no shortcuts to this kind of knowledge. It is the hard-won accomplishment of many years. This is not work for those with a limited attention span, or the easily distracted. I have already said that the painstaking business of counting and cataloging every possible palynomorph salvaged from a piece of clothing or implement can run to hours, days, and weeks. Well, acquiring the broad interdisciplinary knowledge to be able to look at those findings, and have a feeling for the most likely possibilities for where something might have happened, or what a sequence of events might have been, takes much longer. This is a lifetime's endeavor. But when the pieces fit, and an idea of what might have happened suggests itself, the sense of discovery, of a puzzle having been solved, can be immense.

When the girl staggered into her local police station, in the town on the edge of the North Wessex Downs, she was in obvious distress. Her face was smeared and red with tears, her eyes showed the telltale signs of panic and, when the police constable attending the desk escorted her into an interview room to determine the problem, the words came out in a rush. There was a place, only 100 meters from her home, she said, a place with trees and shrubs, a strip of land between two rows of houses. "He forced me down onto the ground and there were woodchips scattered around. He was wearing pajama bottoms, with Snoopy on them, over his jeans."

Sometime later, when I got the call from the police and was asked to attend the case, I heard as much of the story as they knew. The boy and girl had been out together and, on the way home in the dark, instead of leaving her at her gate, he had forced her on another hundred meters or so into this wooded strip of ground. He did not deny having sexual relations with her, but he did deny ever having been to the area where she claimed it had happened. According to the accused, he and the girl had eagerly lain down together on the turf, in the darkness of the public park some 100 meters *before* they reached her home—and it was here that, according to him, they had had consensual sex. Thus, his alibi site and the putative crime scene were about 200 meters apart.

Before I was called, the medical examination of both parties had already been done. The police surgeon had taken swabs from both the alleged victim and the accused, searching for DNA evidence from the tip and shaft of his penis and by swabbing her vagina. The presence of his DNA did not help the police case, however, because the boy had already admitted having sexual intercourse with the girl. Cases like this are the bread and butter of what I do and have been ever since the early "who was telling the truth" case back in Welwyn Garden City.

The scope of a forensic ecologist is wide. If a body is found in an overgrown ditch, I can be called in to examine the site and lay out the possibilities for how the killer may have approached and left the scene of the crime. If a body is found decomposed beyond recognition, we can estimate—sometimes with unnerving accuracy—the length of time that has elapsed between the victim's murder and the discovery of their body. We can locate clandestine burials on small and large scales; we can analyze the contents of your gut to interpret the events that led up to your death; we can identify the residues of poisonous or psychoactive plant materials left behind in cups and other containers, but linking people and places is at the heart of what we do.

Perhaps, in this North Wessex case, I would not be able to confirm whether intercourse had been consensual, but by inspecting the biological trace evidence left in their clothing and footwear, I could probably throw light on where the putative rape had happened. This might tell whose story was false, and whose came more closely to the truth.

It was a golden day in June when I got out of the crime scene manager's car in the North Wessex market town and, together, we made our way to the place where the girl claimed she had been raped. The patch of wooded land was much as she had described it in her statement to the police. The road running alongside it was bordered by a wide strip of herb-rich turf, neatly mown and planted with trees and shrubs. We walked along the path into the little strip of woodland and I immediately took in the large oak tree, as well as the silver birch and the elder bush that grew close to where she said the offense had happened. Under the oak, and immediately alongside the path linking the two rows of houses, was a bare area of scuffed, earthy ground, strewn with woodchips. Between the trees, wherever there was sufficient light, herbs were groping up toward the sky. The houses on the side of the entrance to the path had well-tended gardens, with golden *Laburnum*, early roses, specimen non-native trees, fruit trees, cypress, ivy, and more.

I wandered back to the alleged crime scene. Apart from the woody species, it did not look very promising to me, although I thought it might be richer in trace evidence than at first sight. For one thing it was untidy, with plenty of woody detritus from fallen twigs, and these were bearing mosses and lichens; there were also some etiolated male nettles near the base of the oak. If they had been here together, as the girl had said, they might both have picked up a wide and specific range of trace evidence types, both from this year and residual pollen and other palynomorphs from previous years.

There are few absolutes in biology. Everything is probabilistic whether it concerns human health, the composition of the flora in a rich pasture, the animal and plant species found in any pond, or the growth pattern of daisies in the lawn. The identity and quantity of the various kinds of pollen and spores found in a sample depend on a great many variables, which we call "taphonomic factors." I define palynological taphonomy as "all the factors that determine whether a palynomorph will be found in a specific place at a specific time." It's something that you learn gradually and can never take for granted. It always depends on context and we have to consider its complexities whenever we are interpreting our findings as it is so easy to come to a wrong, potentially damning, conclusion.

Think of what you see in a sunbeam at home. The large numbers of tiny particles could be your own skin flakes, mites, tiny fragments of insects, plants, fungi, or even mineral soil. All these contribute to the "air spora"—a useful term describing all the pollen, spores, and other microscopic entities floating around in the air. Eventually they fall as "pollen rain," creating dust on your mantelpiece or sideboard.

Some plants rely on the wind to disperse their pollen or spores, while others have evolved ways of attracting insects (and even bats or birds) for pollination. Flowers attracting animals are often colored, perfumed, and give nectar while some, which attract flies, even smell like dung. Windblown pollen may travel some distance from the parent plant and, because the process of wind pollination is hit and miss, vast amounts are produced by some species—these are the ones that invariably give us hay fever. Grasses, sedges, oak, hazel, and many others all fall into this group. The flowers are usually green, yellow, or brownish and are insignificant looking. Invariably, they are grouped into inflorescences at the end of stalks, and are like little lambs' tails shaking their pollen into the breeze. Finding windblown pollen on, say, a murder suspect does not necessarily mean that particular plant was present at the crime scene because, depending on the lay of the land, and any barriers such as buildings and vegetation, it may fail to reach the site even though it is capable of traveling considerable distances. A wall or tree trunk can provide a formidable barrier to pollen dispersal.

Just think of the way hazel catkins dance in the wind. These are beautiful, and a symbol of early spring, but each catkin is a mass of male flowers, the structure being so well suited to being flung about in the breeze, hurling out pollen as far as it can go but, even then, most of it lands in a halo around the parent. Just think of a flowering cherry tree when its blossoms fall. They invariably form a pretty pink areola around the trunk, extending out as far as the canopy. The pollen of most plants will form such a ring of varying distance around the parent plant, although some may escape into more turbulent air to be carried up into the air spora.

From a forensic perspective, wind-pollinated plants can often be overrepresented in the samples we take from the crime scene, while insect-pollinated plants can be underrepresented. Their pollen might never reach the air, or never fall much farther than the ground immediately underneath the parent plant, so finding its traces on a suspect or victim's shoes is much rarer, and it can be of considerable significance. Pollen of pine, grass, or hazel all might be overrepresented in the air spora while daisy, clover, buttercup, sloe, and rose might be underrepresented.

Some wind-pollinated plants can be meaningful if their pollen production is moderate—nettle and dock are common, but they are generally more useful forensic markers than grass. Although distributed by wind, in general their pollen grains are fewer and less efficiently dispersed. One gradually gets to know the dispersal dynamics of various species.

There are always exceptions—that is just the nature of ecology—and, in one particular case, the presence of grass pollen, generally not useful, was vital. A colleague of mine in New Zealand, Dallas Mildenhall, was working on a murder case where, as it was revealed later, the offender had dumped his victim in a river and the body had floated downstream from the riverside crime scene. The task was to find the place where she had met her end. After her cardigan was peeled off her rotting corpse, grass leaves and pollen were laboriously collected from it. What was strange was that instead of having a single pore like virtually all other grass and cereal pollen, these grass pollen grains had two pores. This is rare indeed. One occasionally finds an aberrant pollen grain, malformed during development, but this aberration was obvious in all of the grass grains on the victim.

Eventually, the crime scene was found, and the grass identified from its leaves. After testing the various grasses at the site, my fellow palynologist on the other side of the world found at least two plants where all the pollen had two pores. None of us had ever come across this phenomenon before and, with so many palynologists in the world, something like this would have certainly been reported in the literature by now. These grasses with the two-pored pollen had undergone some kind of mutation to produce such abnormal grains, and it is possible that they had, at some time, been treated with some mutagenic chemical, possibly a weed killer. I find that really frightening. I wonder if such a substance could cause mutagenesis in the sex cells of other plants and animals—even human sperm? After all, as bizarre as it seems, we are related to grasses (though distantly) and some of the processes in pollen formation also occur in sperm production. In any event, these mutant grass grains were marvelous identifiers of a place. And, if two-pored grass pollen was found on a suspect, he would have a lot of explaining to do to convince the police that he had not contacted the crime scene. Like many other cases, however, a suspect was never apprehended.

Crime scenes change over time. In one dramatic case, a senior investigating officer in the north of England desperately wanted me to evaluate a crime scene in the brutal murder of a prostitute. It was the weekend, there were no flights, and it would take too long for me to drive all that way north, as this was an urgent case. Thinking out of the box, the senior police officer got in touch with Surrey Police and asked them to bring me in their helicopter. The journey was very quick, but also vastly more interesting, and certainly more fun, than traveling by road, except for having to wear huge safety flying helmets. When we arrived at our destination, my picture was taken as I stepped out of the helicopter. I would have loved to look like a glamorous aviator instead of the bulbous-headed midget that stepped out on that windy tarmac.

A waiting car whisked us to the deposition site, where groups of officers were huddled around, kicking the ground and probably dying to have a smoke. A smiling CSI greeted me.

"I've cut all the grass and bushes down for you, Pat!" he declared. I laughed as I struggled into my Tyvek suit.

"That's a funny one." But, he was not joking. The smile just faded from his face and he stared at me blankly. He had cut back the whole crime scene, contaminating everything, and losing all the evidence I might have found where the body had lain. Not only had the vegetation been cut right back, but it had been dumped, in a complete mishmash, on the only path that might have been taken by the offender. The senior officer came over and was tight-lipped with exasperation. He was as incredulous as I was.

Conversely, here in North Wessex, almost a month had passed since the alleged incident had occurred, and this, in itself, could have posed a conundrum. Nature, after all, is constantly striving: plants continue growing and spreading, changing the look of places beyond all recognition; earthworms keep depositing soil on the surface; plants may die and even disappear.

Thankfully it was high summer and, because I was called fairly soon in this case, the alleged crime scene in the strip of woodland had had little time to change significantly. Unfortunately, this was not the case in the park, just about 200 meters away, where the boy claimed the girl had been so willing. The Council had mown the turf, but the ground was still strewn with grass clippings, with fairly large amounts of clover and hop trefoil, a little yellow-flowered plant related to the clover. The surface of the site that formed the boy's alibi was only represented by the grass cuttings. Although mown, they represented the surface that would have been lain on by the couple. The suspect was very specific in identifying the spot where he claimed they had enjoyed themselves. This meant that I could take my comparator samples from exactly where their bodies would have contacted the ground.

Comparing the two sites with profiles built up from the

clothing of the accused man and his accuser would, in principle, allow me to answer the critical questions: Was the man telling the truth when he said they had lain down together in the park, or should the woman's story—that they had been together in the wooded area between the houses—be believed? But in life, things are rarely either/or scenarios, and so it was here. The question was complicated by the fact that the "Snoopy" pajamas the man had been wearing over his jeans (we never found out why) had been discarded, later to be found hanging over the branches of a *Coto-neaster* shrub at the edge of a nearby garden—perhaps corrupting whatever trace evidence we might derive from them. The situation was complicated even further by the fact that, after the incident, both victim and alleged attacker agreed that they had sat together on the verge outside her home, potentially collecting more contaminating trace evidence there. All of these places would have to be sampled in the hope that we could eliminate them from our investigation. At the two "crime scenes," and all the other places the girl and boy were said to have visited, or contacted, I made species lists of all the plants that I thought could possibly have contributed to the forensic trace evidence.

Here we had two sites, one represented by grass cuttings and the other of soil mixed with woodchips. To get as close to the true picture of place as possible, it is essential to take multiple samples because from experience I know that, in terms of biological content, the surface of the ground, and the soil itself, is extremely patchy over very short distances; one sample would yield only a fragment of the available information. I was very lucky that here the two most important sites were small and well defined, and I could indulge in taking a good complement of surface samples, each one being carefully recorded, with everything photographed as we progressed. The greater the area analyzed, the closer one comes to the truth.

It was very hot, I was sweaty, and the work was incredibly tedious, but that is exactly what forensic science is like most of the time. One also has to keep focused, making sure that nothing has been forgotten. In court, anything overlooked or disregarded could be highlighted, picked on, dissected, and thrown down disdainfully as a challenge.

All the case samples were duly delivered to the laboratory, and the task of getting the pollen grains from the clothing, and out of the comparator samples, began. Such retrieval is a dangerous business as it involves treating the sample with very strong alkalis and acids, including, among others, caustic soda, acetic anhydride, glacial acetic, hydrochloric, sulfuric, and hydrofluoric acids. The latter is horrendously dangerous as the vapor can dissolve your lungs, and the liquid can go right through skin and dissolve bone. It is used in glass etching and, in our work, it is needed to dissolve the quartz out of the soils. Hydrofluoric acid can work its way through glass and metal—it reminds me of the corrosive slime that dripped from the monster that burst out of John Hurt's chest in the film *Alien*, when it went right through the hull of the spaceship.

In hydrofluoric acid, we have the potential for a real-life horror story. Using it requires a high level of protective clothing and a mask, and everything must be done in a fume cupboard. There is a shower nearby, special ointment, and a hotline to the local emergency room of the hospital. You have to be quite robust to do this kind of work and what amuses me is the traditional stereotype of a botanist—tweed skirt, hand lens, stout shoes, and a mild, gentle disposition. In *Jurassic Park*, the "love interest" was an archaeobotanist, and little E.T. was a botanist from outer space, both resourceful and portrayed as mild. I would love to be thought of as being sweet, mild, unchallenging, and gentle. I think I am gentle, but nothing else fits. People with those characteristics make lots of friends but do not do well in court, where the opposition invariably seem like brutal bastards.

The chemical processing did not present any problems and, eventually, long rows of microscope slides were lined up along the workbench. Now the long, grueling counting with the microscope could begin. In every field of view, everything is identified and counted. Sometimes, a palynomorph will elude identification, in which case it is logged as "unknown" and is revisited after counting is complete. If deemed important, long hours can be spent in trying to track it down. Sometimes, I exchange pictures of an unknown with other palynologists in case they know its identity, particularly to colleagues Vaughn Bryant in Texas and Dallas Mildenhall in New Zealand, and they reciprocate when they get stuck too. Sometimes, the palynomorph may elude identification by any of us, in which case it is tallied as an unknown. That is the only way to provide credible data. One cannot afford just to make guesses; the outcome might affect someone's life drastically. If the clothing and footwear profiles from each party resembled the park more than the woodland, then it was reasonable to suppose the boy was telling the truth. On the other hand, if they resembled the woodland profile more, then the evidence would support the girl's testimony.

I started on the comparator samples just to get an idea of the range of taxa I could expect, and to see if they married up with the species lists I had made in the field. That is always of critical interest because it enhances interpretation. The big surprise was that the very open park site, which had a margin of mature trees and few barriers to pollen dispersal, had very little tree pollen in the profile; grasses, nettle, ribwort plantain, other herbs, ferns, and mosses dominated the results. The surprisingly large pollen grains of clover were much in evidence, but I knew that this plant had been growing abundantly in the turf. I had not even noticed the other herbs, so their pollen must have been finding its way into the middle of the park from the edges; the only plants growing in the immediate site were grasses, hop trefoil, plantain, and the clover.

When it came to the enclosed, wooded site, the putative crime scene, it was enclosed in trees and shrubbery. Such vegetation nearly always proves to be a barrier to free pollen flow from outside, but I found twenty-eight different woody plants in the surface soil and litter samples. Pollen from the gardens running down one side of the road had found its way into the site and it was interesting to see how far *Ceanothus* pollen had traveled, and how much cypress pollen had blown onto the surface soil and litter. The only herbaceous plants of note were cow parsley and buttercup, and they were certainly on my list.

It was useful that the pollen profiles from both sites were so different but, although the pollen results were startlingly good, the stark and amazing differences demonstrated by the fungal spore assemblages were even more dramatic. The alleged crime scene revealed a vast range of fungal spores, most of them associated with woodland floors, where dead twigs, branches, rotting leaves, and other debris provide perfect conditions for fungi. Among the twenty-one separate species of fungus identified were the cylindrical, pale brown spores of *Camposporium cambrense*, a common fungus on rotten deciduous tree debris, like birch, holly, and oak; *Brachysporium britannicum*, a common smooth-walled species known from the bark of ash, beech, birch, chestnut, and oak; and, most intriguingly of all, *Clasterosporium flexum*, a fungus whose spores have only been reported six times in Britain, and is found on the decaying leaves of the alien cypress.

Finding such a rarely encountered fungus here might not have been a surprise—there was, after all, a cypress tree growing in a garden immediately alongside the alleged crime scene—but rare palynomorphs provide extraordinarily powerful trace evidence in the world of forensic detection, helping us to make precise correlations more effectively. And nor was this the only rare spore identified here: *Pestalotiopsis funerea*, a parasite of the cypress tree, was there in the assemblage; and so was *Glomus fasciculatum*, a fungus found exclusively in soils and, although extensively found in European woodlands, there have only been four records of this particular species of *Glomus* in Britain. Its rare status was a vital signifier to us. If these spores, which were virtually all characteristic of woody detritus and woodlands, were in the samples from the clothing belonging to either party, we could more confidently place them where the girl said she had been raped.

Crucially, the samples from the park yielded few of these particular fungal spores and, in fact, the ones that dominated our comparator samples from there were much more common: *Epicoccum nigrum*, growing prolifically on the grass cuttings left in the turf; *Cymadothea trifolii*, a fungus specific to clover; and *Melanospora*, a fungus that ordinarily parasitizes other fungi growing on herbaceous plants. The two sites might have had some overlap in terms of their pollen profiles, but their fungi seemed discrete and different.

With rich profiles for our comparator samples established, we could move to the next phase of the investigation: determining which palynomorphs were on the clothing that might persuasively place them in one or the other location. I decided it was necessary to analyze as much of the clothing from both the girl and the boy as possible. I had their footwear, jeans, tops, fleecy jacket, and the bizarre pair of pajama bottoms. If the clothing and footwear profiles from each party resembled the park more than the woodland, then it was reasonable to suppose the man was telling the truth. On the other hand, if they resembled the woodland profile more, then the evidence would support the alleged victim's testimony.

Both sets of clothes had picked up an immense amount of palyniferous material. As the count began, a distinct image formed: the man's jeans, pajama bottoms, and shoes revealed an abundance of taxa we had already identified from the alleged crime scene. And soon, I began to see the same assemblage from the woman's clothing.

The pollen profiles were excitingly comparable. There will never, ever, be an exact match between the samples, but the strength of palynology as a forensic technique is that there are very many markers to take into consideration for interpretation. We are not just looking at one item of trace evidence, as is often the case with fiber trace evidence, or a single gunshot particle. Each palynological profile might contain two hundred to three hundred or more separate units of evidence. Collating and calculating the results from the count data is the most exciting and interesting part of the whole investigation for me.

It is the final picture that makes everything worthwhile and, in this particular case, the results were quite spectacular. The "Snoopy" pajamas yielded a huge amount of pollen from the rose family and this could be explained by them having been flung over the *Cotoneaster* bush which was in full flower at the time. All the trees and shrubs found in the wooded area were also extracted from clothing of both the girl and the boy, especially oak, birch, pine, and elder bush, the four dominant pollen types on the woodland floor.

I was also taken aback at the evidence of contact with the wooded site provided by the fungal spores. Having fungal spores as well as pollen results from each sample meant that we had two distinct classes of forensic evidence, and this made it an exceedingly powerful data set; and the fungal spore results were as spectacular as those of the pollen and plant spores. The botanical and fungal evidence supported each other.

It was intriguing that the absence of certain taxa was as important as the presence. Spores of the fungus *Epicoccum* were overwhelmingly abundant in the grass cuttings, but only one spore was present in the samples from the wooded area. Its virtual absence in the woodland, and very poor representation on the clothing, was significant. With such huge amounts of *Epicoccum* spores in the grass cuttings, neither the girl nor the boy could have failed to pick up large numbers if they had, in truth, lain on the grass. It was the same with the clover pollen. There was a large abundance in the grass cuttings yet, again, the clothing and footwear of both parties yielded only a single grain. This massive difference for clover was as important as the *Epicoccum* spores.

Compared to the park, there was a huge amount of dead, twiggy litter under the trees in the wooded area, and the testament to this habitat was the large number of different fungi which grow on dead wood of various kinds. Further, although microscopic themselves, most of these fungi produce large spores which are not dispersed freely into the air and, if they are picked up, they represent excellent local indicators. The belongings of both the girl and the boy yielded a very wide range of fungal spores that were just not found in the turf of the park and, in particular, the rarely found fungi *Clasterosporium flexum*, so characteristic of dead wood, was an important find. There was more on the girl than the boy, presumably because her body had had greater contact with the ground than his. For these fungal spores to be transferred, there must have been direct contact with the ground in the putative crime scene.

Although there were some traces from other places, and some in common with both sites, there were 115 individual palynomorph taxa to substantiate the girl's story. In the knowledge that, in my twenty-five years as a forensic palynologist, I have never had two locations yield the same palynological profile, I must ask what is the likelihood of two people achieving these profiles by chance from randomly chosen sites? The pollen and fungal spore profile obtained from the girl's clothing resembled that from all the boy's clothing and footwear closely. Then, both sets of clothing were much more like the woodland profile than that of the park. This merely confirmed the story that they had both lain together in the wooded strip of land, and the putative crime scene then became a real one.

We do not need perfection to have confidence in our findings. Nature has given us a messy, imperfect world and rarely, if ever, do things match exactly. Yet there was no denying the strength of the correlation we had uncovered here. The assemblage on the suspect's and alleged victim's clothing closely resembled each other and, taken together, they closely resembled the samples we had taken from the woodland—the place where the girl had claimed rape.

I cast myself back to that space between the trees, overlooked by the manicured middle-class gardens where alien species of tree and shrub were in full flower. Of these two places, this, I was sure, was where the two had wrestled on the ground together; it was here, in the shadow of these trees, picking up the pollen and spores from the debris littered on the ground, that they had struggled. We might not have had witnesses. We might not have had testimony we could rely on. In another era, this might have been a cut-and-dried "him versus her" case, the kind that will almost always fail the "burden of proof" test that is so critical to our courts. But what we did have was a picture, built up from pollen and spores, of where these two young people had been, of the unseen nuances of their surroundings that their clothes and bodies had rubbed up against. Nature had left its imprint on them and that was enough, in the eyes of the law, to corroborate one story beyond reasonable doubt, to vindicate a victim and expose an accused man as the rapist he was.

One thing that needs mentioning was that one of the first tasks in this whole case was to analyze the swabs taken from the genitalia of the two litigants. No particulates were retrieved from the male, but the swabs of the girl's vagina yielded two grains of oak pollen and one of grass. It is obvious to anyone with an imagination that the boy's hands might have transferred the pollen to his penis then secondarily transferred it to the girl. Or, of course, his penis might have touched the ground, and then there could have been a direct introduction. The tree under which the attack had taken place was an oak, and how could that oak and grass pollen have reached the intimacy of the girl's body unless it had been put there?

Once the young man's attorney had had time to digest our evidence, the boy quickly confessed. This saved the country a lot of money, as court cases cost a fortune. I am glad to say that a fairly large number of our cases have resulted in confessions, so we have saved the taxpayer quite a lot. But although it is always important to be careful of confessions in case they were obtained under coercion or other illegal means, this putative rape was certainly recorded as an actual one, and the young man was given a custodial sentence commensurate with the harm he had done.

CHAPTER 7

A Spider's Web

Pollen and spores are robust and, where bacteria and fungi are kept at bay, they can endure for thousands of years, even millions of years in some rocks. In archeology and ecology, this makes them of great value for reconstructing past environments and demonstrating landscape change. Plants are also pivotal in human history, and their remains are testament to the ways they have been exploited over millennia.

On one occasion I was briefly baffled by a case brought to me by the police from a mining town in Yorkshire. Two boys had found a heavy, new sports bag lying in the road. It was so heavy that it must contain a lot of equipment, and they carried it home, full of excitement. They could not wait to see what was inside, but the fright they received served them right; what was revealed by the unzipping would have unnerved the toughest of men. A mummified corpse, as alarming as anything they had seen in horror films, was inside, and the boys' screams and yells brought their parents running, and soon telephoning the police. No one had a clue about the identity of this dead man.

Except for the lower legs and feet, which were inside a loosely wrapped plastic trash bag, the whole body was tightly encased in cling-film plastic. A solitary yellow sycamore leaf was stuck to the thigh and, from the shins down, the skin was covered by a black, sooty material. The two officers who brought me the sample said that the corpse stank and the smell was very likely that of the black stuff. They agreed with me that it was a chemical smell and probably some kind of engine oil. I quickly made a temporary slide preparation of the "soot," and the view under the microscope left nothing to doubt; it was a thick mass of fungal spores. This presented us with quite a puzzle. I realized that the looseness of the wrappings around the lower legs and feet meant that they were exposed to the air, whereas the cling film prevented oxygen from reaching the rest of the body. Nearly all fungi (though as recently found, not quite all) are like us in that they need oxygen to live and grow. This meant that while the rest of the body was the brown of mummification, the feet were as black as soot, and this soot consisted of the spores of an actively growing fungus.

Everyone was interested in the leaf and, after some effort, I was able to obtain quite a few pollen grains from its surface. Again, the picture of place was an unkempt garden and, strangely, there was plenty of rose-type pollen, some *Clematis*, and some pollen of sycamore, pine, and birch. These were all hints of the place the leaf had come from. A postmortem had established that the man had been stabbed, but it was impossible to determine how long he had been dead. What was curious, though, was the sand that was stuck to the skin on the back and front of the torso and in his hair. The young police officers were flummoxed, although it was obvious that they were enjoying the variety offered by this case, their first encounter with a murder.

As it turned out, the victim was identified by a sheer fluke; the facial reconstruction eventually came through and "wanted person" posters of the face were distributed all over this part of Yorkshire. Bizarrely, it was one of the police vehicle mechanics, working in the depot attached to the police station, who recognized the face as belonging to a Yemeni immigrant who lived not far from the police station. The extraordinary twist in the story was that the victim had worked in the same depot before he had gone missing. His home and family were quickly identified and it was of interest to the police that his house had been put up for sale.

When I got to the terraced red-brick Victorian house, I was struck by the mutilated remains of what had been a huge rosebush immediately outside the back door, which opened onto a tiny, concrete backyard. The remains of an old *Clematis* plant still straggled on the fence next to the rose. The rest of the garden had been stripped and was weirdly empty; somehow, the topsoil had all been removed. But there was a sycamore tree at the end, overhanging the garage and the next-door garden.

The pollen profile suggested that when the killers had wrapped the body in plastic cling film, they had laid it in the backyard, next to the rosebush and *Clematis*. There was a great deal of organic, decomposing plant litter on the concrete and this had probably collected rose and *Clematis* pollen; this could have happened when the bush was chopped down and residual pollen was recirculated.

The yellowness of the sycamore leaf yielded information too.

In fact, from that single leaf, I was able to build up a picture of the main features of the part of the garden near the house, and the possible circumstances around the man's death. If it had been pulled from the tree when green, I would have expected it to keep its greenness since chlorophyll does not readily break down when leaves are detached prematurely from the parent plant. But, an autumn leaf that has already yellowed would have floated from the tree naturally and could have blown up onto the backyard from the bottom of the garden. This gave a hint that the corpse had probably been wrapped up in the latter part of the year rather than in spring or summer. But sycamore leaves decompose quite quickly and might have contributed to the organic mush on the concrete if we were looking at late autumn or early winter. What had been picked up by the body had been an early autumnal leaf, and it had been protected from breakdown by the very circumstance that prevented fungi from growing on the body other than the feet; a lack of oxygen, because of the tight plastic overcoat encasing the corpse, had prevented decomposition of that important piece of evidence. Indeed, that single leaf had preserved the picture of the garden to tell us part of the story but there was plenty of other evidence to implicate the victim's son and grandson in the crime.

The cellar of the house revealed a burial place, as well as the source of the sand particles that had been stuck to the victim's skin. At the bottom of the cellar steps, officers found a brick-built cell, freshly painted in green gloss paint and filled full of sand. Analysis showed that it was builders' sand, and that it was heavily contaminated with diesel oil. We were astounded to learn that during the Gulf War, the Yemenis routinely doused the dead with engine oil and then buried them in the sand. This was certainly not a Yorkshire tradition and long police interviews with the perpetrators revealed that the son and grandson had copied what they had witnessed in the Gulf War. The grandfather had been a domineering and cruel old man. They just could not tolerate him anymore and, in an outburst of anger, resorted to drastic action when, as punishment for some trivial misdemeanor, the old man had put his kukri knife in the fire and deliberately burned his grandson's leg with the blade. They had grabbed the knife from him and stabbed him with it.

Early in the investigation, I was asked to find out as much as I could about the house, the people and their habits. There was a coal shaft leading straight from a trapdoor on the outside pavement into the coal cellar, and it was impossible for me to resist taking a sample of the coal dust from the shaft. The contents of that sample were difficult to interpret, so much so that I got a headache, which only lifted when I realized what I had found: a rich profile representing a hay meadow, with pollen that was so well preserved that it looked as though it had been shed that day. But then a moment of epiphany: I remembered that, when I was young, people still used horses and carts. These would once have plied these terraced streets daily and, when the coalman stopped to shovel coal into the cellar, his horse would have had a break too. From my childhood in Wales, I well remember dollops of steaming manure, at intervals along the street, dropped by the various tradesmen's horses. It was precious stuff and keen gardeners used to rush out to shovel it into buckets. Of course that was the source of the hay meadow in the cellar.

Horse dung is very friable when it dries, and it blows around, into gutters and any nook and cranny. Generations of dung debris, with its pollen load, had found its way into the coal shaft, and the sulfurous coal dust had been so acidic that all microbial activity was prevented such that the pollen was beautifully preserved. It had survived the extreme chemistry of the horse gut and had lain in that shaft probably since Victorian times, but certainly since at least the 1940s or even early 1950s.

Such a discovery has helped me better understand weird pollen profiles ever since. I have learned to envisage a site as it had been, or is being, used and I have since found hay meadow pollen deep in woodland, away from any farmed fields—again the result of dry horse manure blowing in from bridle paths. It is no surprise, then, that prehistoric pollen has been extracted from the dung of wooly mammoth, last known to roam Earth some ten thousand years ago, and from the gut of the Iceman found in 1991, a Neolithic hunter who roamed the Alps over five thousand years ago, the arrow that killed him still embedded in his shoulder blade. Often, these pollen and spores from guts are very well preserved indeed.

With pollen grains surviving for millennia like this, it is perhaps no surprise that the clothes of the alleged victim and perpetrator in the North Wessex rape case had proved so important. But even when a person has been sufficiently forensically aware to destroy their clothes, or swap them with someone else's, trace evidence is not necessarily lost. Alive or dead, most of us, my dear husband ex-cepted, have a natural attribute that can trap palynomorphs. Our hair follows us wherever we go and, whether we use hair spray, gel, or other products, pollen and spores will stick to it tenaciously by electrostatic forces, and this leads me to another case entirely.

The girl had been missing for almost a year when, in the dying days of summer 2001, she was discovered in an excavated depression on the borders of a Yorkshire forestry nursery. She was still wrapped in the duvet which her killer had hastily put around her body. Not yet fifteen years old when she had vanished, her disappearance on the way home after a shopping trip with friends had sparked one of the largest missing person's operations in the history of Yorkshire policing. Two hundred officers and hundreds of volunteers had fanned out through the streets and along the bus route she took home, knocking on thousands of doors, searching 800 houses, sheds, garages, and outbuildings. Search warrants were issued, 140 men with past convictions investigated, collections of household waste curtailed while refuse sites were searched—and a local benefactor even offered a £10,000 reward for information leading to her return. But none of it mattered: she would never return home.

It was a dog walker who found her, as it so often is. Her grave lay only 100 meters away from the spot where another murdered girl had been discovered some ten years before. Inside a floral polycotton duvet cover, her body was swaddled in several green plastic trash bags; a black trash bag was over her head, secured by a dog collar around her neck. At the burial site, a mixture of native hardwood species had been planted along the adjacent road. This is commonly done by the Forestry Commission to give the impression that their alien nurseries are more diverse than they really are. Emerging from the woodland edge and toward the road, there was an open area consisting of acid grassland, with occasional patches of heather, bilberry, and bracken fern. This kind of terrain is common in Yorkshire and other parts of the UK where the Forestry Commission, and private landowners, spread these dark, brooding woodlands all over our hillsides. They are shady and unkempt inside, and the trees are invariably regimented into straight lines; they are deeply forbidding to our native wildlife. I always view them as blankets of doom and gloom, and can hardly bring myself to grace them with the title "woodland."

When I arrived at the burial site, some police officers and a couple of suited-and-booted scientists were busying themselves, but the scene was quiet and hushed. I was not very pleased to see the forensic archeologist already into deep excavation, having paid no attention to preserving the area immediately around the grave. This meant that critically important places for me to obtain my surface samples would have been destroyed by being contaminated by the grave-fill. I busied myself taking uncontaminated surface samples as close to the grave as I could, and surveying the vegetation around and in the vicinity of the grave, and along the offender's putative approach path. I made a comprehensive species list so that, if it was necessary, I could compare this woodland's botanical profile with any that I obtained from the offender's belongings, assuming that the police would manage to arrest one.

There was a small silver lining among this grim, desolate scene: the way this poor little girl had been wrapped up meant that she had not been exposed to the grave soil or surrounding vegetation. Whatever trace evidence I might retrieve from her would not represent the woodland of her burial but would represent the last place she had made contact with the "outside world." That might lead us directly to her killer.

It was with this in mind that I arrived at Leeds General Infirmary a month after the girl was brought up out of the woodland. Her body had been made ready on the stainless steel mortuary table and my first job was to wash as much trace evidence as I could from her detached scalp.

Hair, and the same goes for fur and feathers, is amazing stuff. It is made of the highly resistant protein keratin, which also makes up nails, hooves, and claws. It is strong and durable. The only other biological matter that comes close to it for toughness is chitin, the principle component of the shells of crabs, the outside skeletons of insects, and the walls of fungal cells. Natural fiber does not get much tougher than hair and its durability is a great boon for those of us looking for particulates stuck to its surface. I have found that if hair contacts a surface covered with particles such as pollen, any kind of spore and even minerals, they will quickly be transferred to any strands coming in contact.

Each hair is made up of layers: first, an innermost medulla, which is present in only the thickest types of hair; then a cortex and, finally, an outer cuticle, consisting of overlapping scales that gradually erode as the hair grows old. Because hair exhibits various static electric forces, it actively attracts particulates—and this means that it can, under the right circumstances, act like a spider's web, trapping palynomorphs that we can later recover. Pollen and spores can be held on hair and fur for incredibly long periods. In archeology, they might be recovered millennia later and used to help us reconstitute an idea of the prehistoric landscape. In a forensic context, pollen and spores might be retained on hair indefinitely. Before, during, and after death, hair will draw palynomorphs to it, so I can retrieve evidence from it. From a victim's hair, I have often been able to envisage the kind of place where the body has lain, even when it has been in more than one.

Sometimes, the hair of a murder victim is clean and well groomed, but not always. I often have to work on hair matted with blood, other body fluids, and the slime of decomposition; it may be coated with dirt, soil, and other materials. Even a week or so after death, the scalp can become detached from the skull and, with the skin tissue putrefying faster than hair decays, it is not unusual for a victim's hair to be found some distance from the corpse itself. Indeed, at crime scenes where corpses are exposed on the surface rather than buried, hair can become spread out across the surrounding ground. I have seen birds gathering it; it is, after all, excellent nesting material.

One of the reasons for the monthlong delay in my part of the investigation was that the state of preservation of the victim was so good it was thought impossible for her to have been buried for the whole eight-month period of her disappearance. The pathologist felt she might have been kept in a freezer, or somewhere else very cold, and then buried at the offender's convenience. This meant that she could have been in her grave for a relatively short period. Of course, one had no idea of the length of the freezing period even if, indeed, she had been frozen. The police employed experts from the frozen food industry, and the state of muscle in frozen meat was investigated. However, as in so much of forensic work, observations and experiments are not strictly scientific because one can never replicate the original conditions to test one's model or hypothesis. Everything must be approximate, but it is the best that can be done under current legislation, and it is certainly worth attempting more creative ways of testing one's ideas.

"What we need, Pat," the detectives had said to me, "is some idea of the place she's been kept all this time . . ."

It is at moments like these that I become distinctly aware of what a huge responsibility my work can be. The police, the girl's parents, the press, and everyone else all want to know the same thing, and I was trying to provide answers. The weight of it can be oppressive.

When I was called to the mortuary, the air smelled strongly of the disinfectant from the footbaths at the entrance, and it was excessively bright, the light harshly reflecting from the metal surfaces. I had already changed into my blue scrubs in the changing room and donned the customary white Wellingtons that, of course, did not fit. My feet are as small as a child's, and I usually have to scuff them along the floor to the metal table where the business is done. I quickly set up a bench with my essential bits of equipment: stainless steel bowl and jug, scalpel, forceps, sample bottles, labels, bottle of bleach, and medicated shampoo. My job was to extract any evidence from the corpse's hair, as well as her nasal passages, mouth, and skin. By analyzing particulates or fragments of vegetation, or anything else I could find that might have been missed, I might be able to build up a picture of what happened around the time the girl died and, importantly, the kind of place where this happened.

I was surprised to discover that I needed to remove so much plant material from the girl's body and, at first, thought it might be useful to the case; but I soon realized that whoever had removed the girl's body from the trash bags had not been careful enough, and some of the grave-fill had landed on her skin as the bags were being opened. Once again, botanical contamination had not even entered the head of the person unwrapping the body. I was only glad that nothing had fallen on her hair—otherwise I would have had to try to separate the palynological profile of the burial site from the one we were seeking.

Lying cold on stainless steel, the girl was in a fairly advanced state of decomposition, even though she was too well preserved to have been at higher temperatures for eight months. Although the body had not decayed beyond recognition, it was still actively decomposing, and the stench as I approached was overpowering. I nearly retched, but managed to internalize the urge and get on with the job. The smell of a decomposing corpse is due to a complex mixture of odors that are the by-products of autolysis and bacterial enzyme action; it is a truly disgusting smell, and it can change over time as decomposition proceeds. When your heart stops beating, your body does not die immediately. Certainly, your brain stops functioning, but it takes a little while for the cells to fail, and the body probably fades rather gradually, although it is thought that some parts start decomposing within about four minutes. Certainly, the melanocytes in the skin can still function for at least eighteen hours after death. I remember one case where a leaf had floated down from the woodland canopy onto the leg of the young woman's naked, pale corpse spread-eagle in a clearing. I picked it from her thigh and was very surprised indeed to see the white leaf shape left on the skin. She was fair and her skin seemingly white, but the skin all around the leaf must have become very slightly tanned as she lay exposed in the dappled light of the woodland.

The body in front of me stank of cheese and feces. Obviously, the butyric acid bacteria were very active at this particular stage of decomposition. The skin was also slimy, probably teeming with decomposer bacteria and yeasts. Because the scalp was already detached, I could just put it into a deep stainless steel jug and agitate it vigorously in a dilute solution of my "friend and companion," medicated shampoo. Once I felt I had treated the hair well enough, and the water was sufficiently cloudy, I poured the washings into plastic universal bottles, labeled each with detailed descriptions, and put them to one side to take back to the laboratory in London. To make sure I was maximizing retrieval, I then rinsed the hair with as little water as I could get away with and put the rinsings into another set of bottles. All this dirty water was one sample and I would eventually mix the contents of the bottles together, and then split the mixture into two parts. If one were lost accidentally, I would still have the other portion.

Throughout my time in the mortuary—while I also looked

into the girl's lips, gums, and nasal cavities—the mortuary technicians were friendly and helpful. I was thankful for this because I
do not always get wholehearted acceptance, especially by pathologists. Pathologists are a mixed bunch and many of them suffer
from the "deity delusion." They are gods in their own mortuaries
and, although there are good ones out there, some really seem to be
offended by any other specialist coming in with what they think
are wacky ideas.

While I was preparing my samples, a voice called out across
the mortuary.

"I'm sure you'd like some lunch, wouldn't you, Pat?" I looked
around and there was the friendly face of the Senior Investigating
Officer who had called me into the case.

"Oh, yes, please," I answered.

When we got to the staff refectory, there was little choice because I had taken so long in sampling the body.

"What would you like, Pat?"

I was too tired to bother and, to be honest, I really did not feel
all that hungry anyway. But, I had a long drive back south and
needed to have something to keep me going.

"Why don't you choose for me?" I said. "Anything that doesn't
have meat, please . . ."

Moments later, he returned with a laden tray and started placing
the plates on the table. I looked, smelled, and immediately felt nau-
seous. It was cauliflower cheese, a dish I usually relish, but it smelled
of butyric acid, with a whiff of hydrogen sulfide; in short, it smelled
of the corpse. The color was of flesh in livor mortis, with slightly gray
tinges at the edges of what looked like bits of brain. Well, of course it
did—butyric acid came from the cheese and the sulfur compound
from the cauliflower. The cabbage family, which includes the cauli-

flower, produces many sulfur compounds and I suppose this is why some people hate cabbage, cauliflower, and sprouts. Butyric acid is formed by bacterial fermentation, and the bacteria involved in cheese-making are the same ones involved in both corpse decomposition and the smell of sweaty feet. I tried to be objective but, as I tried to eat my meal, it was coming up as fast as it was going down. I mentally slapped my own face and could imagine my grandmother saying, "Stop whingeing and get on with it!" So that's what I did.

Clutching my precious bag of samples and equipment, I set off on the long drive south. It took over seven hours because of all sorts of hold-ups on the motorways. I eventually got home, flopped down in front of the TV at about 1:00 a.m. with my darling Mickey, my one-eyed, silky Burmese cat, on my lap. I was next conscious at 4:30 a.m. with a crick in my neck, and Mickey's whiskers tickling my cheek. The discordant music of some ridiculous horror film was screeching away on the TV. I switched it off, Mickey still in my arms, and we both went up to bed, not waking until about 10:00 a.m. the next morning.

I had discovered that hair was a marvelous source of pollen and spores just by trial and error, and had been amazed at its power of attracting microscopic particles. The research of one of my master's students had shown us that the presence of styling products or hair spray makes no difference. Pollen and spores will stick to clean, coated, or dirty hair. In postmortem examinations, I had seen pathologists perfunctorily passing a comb through a victim's hair and, at the time, always thought that was a bit useless for retrieving evidence. The whole head of hair needs to be sampled, and combing a small portion of a scalp would not even get the larger particles of plant material efficiently. It is amazing how protocols and methods can become so entrenched that their execution becomes a box-checking exercise rather than a meaningful investigation. I never take anything for granted and, because I am rather curious, I always want to go that little bit further. This led to some of my best results in murder cases.

I remember on one occasion the victim was a black Afro-Caribbean lady, dressed in an expensive, but uncoordinated, outfit. I had never even touched Afro-Caribbean hair in my life and, in the mortuary, I tried to wash the victim's hair in a bowl as I had always done. But it would just not get wet. It was a little like the soft hairy leaf of the lady's mantle plant; the very fine hairs repel water so effectively that it just sits like little diamond droplets on the surface. The pathologist, one of the most inventive ones I had ever met, came up with a solution. By cutting around her neck and throat, he just slipped off the whole face and scalp. I was left speechless. Here was a glove puppet, and a corpse devoid of face and scalp. I put my gloved hand inside the "face glove" and agitated it thoroughly in the bowl of hot detergent solution. I managed to get a good sample. We then replaced the scalp and the "face glove," and you would never have known that she had even been touched. The eyes fitted back perfectly into the eyelids, and she did not look any different from before. I now realize how a plastic surgeon must regard the face. It is just a thin, flexible covering on top of muscle and bone. What shook me was how easily and neatly a plastic surgeon would be able to cut, shift around, and even remove skin to suit the needs of his patient.

Back in the lab in London, I set about processing our samples, centrifuging them all, and discarding each "supernatant"—the fluid on the top of the little plug in the bottom of each tube.

Processing eight samples can take all day and involves digesting the background of unwanted cellulose, other polymers, and silica, with a series of strong alkalis and viciously strong acids, including hydrofluoric acid. Eventually, I had a long line of mounted samples on my bench and I got on with the microscopical analysis. The telephone rang shrilly at the side of my head and made me jump.

"Anything yet, Pat?" came the policeman's voice on the line.

Yes, there was. Already it was coming into view. I could see yet another very neglected, run-down garden, but this time there had been a bonfire. Both conifers and hardwoods had been burned, their specific kinds of wood cells being testament to that: tracheids in the case of the conifers and wide xylem vessels from the hardwoods. Their anatomy was beautifully preserved as charcoal, but there was also a great deal of black, burned, amorphous material as well as angular and rounded silica grains—sandy grit. There was too much grit and charred debris in the hair for it to have had brief contact. It seemed to have been lying either very close to, or even in, bonfire ash.

The place jumped out at me. The girl had been placed near a privet hedge that was probably very neglected. It must have flowered quite prolifically to account for the relatively high levels of its pollen in the victim's hair. This meant it had probably not been trimmed for a long time; possibly the bushes were large and the hedge tall. It also meant that she had lain close to the hedge—privet is insect-pollinated and does not produce vast amounts of pollen, and the victim's hair yielded more of this than could be expected. Although poplar is wind-pollinated, and the pollen can travel considerable distances, its abundance probably meant that there was at least one tree growing close to the privet. It produces catkins and abundant pollen, but the grains are thin-walled,

138

spherical little blobs, with minute scabs on their surfaces, and it is not very robust so probably breaks down quickly. I have only rarely had cases where I even found this pollen, let alone an abundance of it. To me, this was a good marker.

Other types that might have been important were elder, beech, and *Prunus*-type: perhaps plum, cherry, damson, or sloe. If the pollen were of sloe, this could be a wild hedge—but, when all the herbs started filling the picture that was unfolding in my mind's eye, it was obvious to me that this place was a neglected garden. First of all, there were twenty-four grains in my pollen count that I did not recognize at all. They did not key out in my identification keys, so they were likely to be non-native cultivated species. I did not want to waste hours on definitive identification; it can sometimes take many hours of working with reference material and literature, and the police wanted quick answers. At this point, it was enough to be able to say that this was a garden, or possibly a park. I decided to be pragmatic and concentrate on pollen and spores that were immediately recognizable; I could always go back to the difficult ones later. The whole palynological profile was dominated by all sorts of weeds, many of which are characteristic of open ground and disturbed soils—dandelion-like plants, goosefoot, nettle, shepherd's purse, cleavers, and many others. There were also buckler ferns and even *Sphagnum* moss.

I have long ceased to be surprised by finding bog and moorland plants like *Sphagnum* in urban settings. When you think how much peat has been cut from our uplands to satisfy the horticultural business—from Scotland, from the Pennines and the bare but beautiful Irish landscapes—you might imagine the vast quantities of their spores and pollen sitting in flowerpots and urban soil wherever you get keen gardeners. A surprise was that the girl's hair yielded a large amount of fungal hyphae and huge numbers of fungal spores. It was unlikely that they were growing on the hair itself because very few fungi have the enzymes to break down the keratin of hair, nails, and feathers. Those causing ringworm and toenail infections are soil fungi, and are capable of using hair as food, but the fungi in the girl's hair were more likely to be those normally growing on plant debris, and the hair had probably picked them up through direct contact with plant litter.

I had also found cereal pollen in the hair. Perhaps this meant that the gardener had, at some time, grown strawberries or rhubarb? This might imply straw and horse dung. Both are used in cultivating these plants, and even well-rotted horse manure usually contains bedding straw from the stables. I gave the investigators a description of the kind of place the poor girl had lain before being buried and, although it was limited, at least it told the police that they were looking for a domestic scenario rather than some wild place. They were also seeking someone who might have a somewhat "casual" attitude to their home and garden.

As early as my first case in Hertfordshire, I had realized the necessity of eliminating any irrelevant, or alibi, sites and I knew I had to do the same thing here. In the case of the Chinese Triad and the murder of the hog-tied man who had been abducted on his wedding day, I had to make certain that the pollen I had recovered from the suspects' car really had come from the hedgerow where the victim had been dumped. So accompanied by bemused police officers, I made trips to the known haunts of the victim and the accused in the East End of London. I made species lists and collected samples of the ground likely to have been contacted by the offenders in as many areas connected with the suspects as I could. I tried my best to eliminate as many places as possible as being the source of the pollen in the vehicle that had been seized from the suspects. Our mission that day may have been successful, but it sticks in my mind now for altogether different reasons.

Walking along, my head buried in my notebook with an enormous policeman flanking me on either side, I did not understand that anything was wrong until one of the policemen said softly, "We have to go now, Pat."

In those days, I was green; I was taking my first steps into detective work and perhaps I had spent too long in the confident safety of university life to even think about the world to which I was being introduced. I was not ready to go, I protested. There were other places we might visit; other stands of vegetation to list. This was my first attempt at police work and I wanted, above all else, to be thorough. And yet . . .

"No, Pat," the policeman said, more severely, "we have to go now." It was only then that I looked up. I was with plainclothes detectives but, somehow, word had got around; somehow it was obvious who we were. Frankly, we must have looked suspicious— and certainly comical: two huge men in dark trousers and smart white shirts, and a little lady in the middle scribbling in a notebook. I doubt that the residents would have seen anything like that every day. Now figures lurked on the corner of every street. Eyes watched us from the pavement, from the crossroads, from the end of somebody's front yard. It was the first time that I properly understood: in police work, you have to be careful about where you go. And yet go you must. The cost of a wrong interpretation is severe, but not being able to back up your findings with comparators is more severe still. Prosecution cases can collapse for the lack of supporting evidence for claims made in reports. It pays to be thorough.

Back to the case of the girl in the duvet cover, I had to exclude all irrelevant places for elimination purposes. I had already taken samples from the front and back gardens of her home, as well as the last place she was seen. I had to be sure that the palynomorphs in her hair truly represented the place or places she had encountered after her abduction. I had also made long lists of plants, both in these gardens and all those in the close vicinity of her home. Defense attorneys are clever and all they have to do to diminish a prosecution argument in court is suggest that anything from the body could have been picked up at some place, or places, other than those associated with the defendant. Very early in every case, I start thinking of the courtroom, envisaging the questions a defense attorney might hurl at me as they do their best to destroy my evidence. I then try to answer those questions myself and mentally plug every little gap in my investigative protocol.

While I was working away on the pollen, the police were doing inspired detective work. With the story of the girl's disappearance dominating the news, her face emblazoned on every milk carton sold by the supermarket chain Iceland, two people contacted the police separately implicating the same man. They had both met him through a lonely hearts column, and he lived on the very same housing estate as the girl's family. This suspect was just another ordinary man, according to his neighbors. He sold pet food for a living, regularly poached in the same woods where the girl's body had been found, and kept himself to himself—but, as is so often the case, a banal exterior masked an altogether different kind of character. According to his past girlfriends, he had a sexual need for bondage, tying up his lovers and locking them in cupboards. To her horror, he had confessed to one of his past lovers that he wanted to bind her daughter in cable ties and have intercourse with her.

All of this put this man firmly in the spotlight, but it was not yet enough for a police warrant to be issued. And, as detectives worked earnestly to uncover more than hearsay against him, there was still much I could do to advance the case. It was important to link the girl with this man, or link her with a place associated with him. Before 2008, the legal system in England and Wales stated that a defendant must be considered innocent until proven guilty "beyond reasonable doubt." It is entirely the responsibility of the prosecution to prove guilt. This requirement still stands, but now the judge must instruct the jury that they "must be satisfied that they are sure the defendant is guilty." The concept of "beyond reasonable doubt" is open to interpretation and, where my kind of trace evidence is involved, it is all too easy for a defense attorney to suggest that any palynomorph profile could have been acquired from some other place. It is interesting that nearly every time I have given evidence in some serious case or other, it becomes blatantly obvious that the attorney knows little about science and nothing about botany. Some have their juniors burning the midnight oil trying to construct questions that will catch me out but, because they just do not understand the implication of their questions, they are invariably easy to answer.*

---

* I always see my evidence as being the ammunition. I provide the bullets to the attorney—he is the weapon, who aims and delivers them to the other side. If he is not a good shot, he fails his client, and he fails the court. In my experience, there are plenty of bad shots in the legal world. I have come across only one defense attorney who gave me a truly difficult time when I was standing for the prosecution. We had worked together on a very high-profile murder of two pretty little girls from East Anglia, and I had spent long periods with him in chambers teaching him the strengths and weaknesses of biological evidence. After that session in the Old Bailey, I had not met him again until a case in Ipswich for the murder of a woman. That day, I deeply regretted the thoroughness of my teaching, and my introducing this sharp-as-a-razor man into my world. My encounter with that attorney, Karim Khalil, with whom I am now on good terms, is a story in itself. I admire him (I did so even when I hated him for giving me a hard time), and think that he well deserves his appointment as a recorder and part-time judge at the Old Bailey.

At last, the police had a breakthrough that allowed them to exercise a search warrant at the house of the pet food seller with a penchant for bondage. The leather dog collar which had fastened the black trash bag around the girl's neck was found to have been manufactured by a company based in Nottingham. This company supplied more than two hundred retailers—and by painstakingly checking these one by one, detectives discovered that a mail-order company, based in Liverpool, had made three sales to addresses in the area where the murder had occurred. One of those sales was registered to this man's address. This was the justification the police needed to enter his home and the garden behind it.

I already knew what the police would find when they entered the garden. I had seen it in my mind's eye. And when the police took me there, I recognized it. The garden was small, typical of the area, but what struck me immediately was the large damson tree with its canopy overhanging a great deal of the path. So this was where the plum/damson/cherry pollen had come from. Just to the left of the back door was a solitary hutch and I noticed movement. Oh, a lovely little ferret stared back at me with its kitten-like face pressed up against the mesh. My main concern now was the welfare of this poor, caged, little innocent who had been badly neglected during the whole operation. I refused to do anything else until I got a firm promise that the ferret was a priority. One thing I did notice immediately was the straw in the hutch—cereal pollen? In fact, there was quite a lot of straw strewn around the garden. The man had bred dogs and, to the left-hand side of the garden, was a succession of brick-built kennels, now derelict and fallen into disrepair, but they still had old straw scattered in them. Just a bit farther on there were the remains of a bonfire and, on a little farther still, was the garden boundary. It was separated from neighboring properties by a huge, out-of-control privet hedge and hanging over this were the lower branches of a poplar tree and the branches of an elder bush pushing through a gap. There were abandoned, scruffy flower beds that had most likely been planted with a range of garden flowers. Presumably, these would have accounted for the pollen I could not immediately identify, but that was of little importance now. It was enough to know that there was evidence of former cultivated garden plants.

There were the remains of another bonfire toward the fence, on the opposite side of the garden to the first one, but I knew the girl's hair had not been in contact with its ash. The vegetation at that side of the garden was decidedly different from the plants represented in her hair. There were also brambles rearing up and over the rubbish piled up on that side, and the whole area was being strangled by bindweed, with some stunted willowherbs poking through the tangled mass. If the girl had been close to them, there was a likelihood that they would be represented in her hair, but they were not. Those in my profile from the hair and the duvet cover were all around the first bonfire and the bottom of the privet hedge, under the poplar and elder. The abundant but stunted weeds, struggling in the compacted soil of the yard, were the ones in my pollen and spore profile. The most abundant were goosefoot, nettle, clover, cat's ear, and sow thistle, the latter two accounting for my dandelion-type pollen. This side of the yard also had many other weeds, but they were less frequent over by the other bonfire remains because it was so overwhelmed and overshadowed by bindweed and bramble.

There was little doubt in my mind that the girl and the duvet had been lying in this backyard for some time before she was buried. Her hair must have been falling loosely onto the ground. The comparator samples I subsequently took from the yard simply confirmed what our previous analysis had already demonstrated. There was a high likelihood that this was the garden and the specific part of that garden where the girl had lain.

As I have already said, there are always anomalies. Perfection does not exist, in this or any other world. The garden had shown evidence of plants not represented in the profile—but this is to be expected, especially of wind-pollinated taxa. Nothing is ever 100 percent accurate in forensic ecology, but, combined with the dog collar, the evidence I had provided was strong enough for the killer to confess his guilt. It spared his victim's family the agonies of criminal process and reliving her last moments in court, but I was confident that, even if he had claimed innocence, the palynological contribution was strong enough to move any court case toward a conviction.

As it was, his story transformed several times; first, it was an accident that he was compelled to cover up; then it was an impulse killing that not even he understood. First, he had buried her in the forest immediately after having killed her; but then he came out with a version that felt perilously close to the truth, and one that the palynomorphs had been telling us all along: after the murder, and with no place else to hide the body, he had wrapped it in the trash bag and the cotton cover, and hidden it beneath pallets in his back garden. Her hair must have spilled out onto the ground, picking up all the pollen I had found in it. Perhaps that drove him to put her head in a trash bag and secure it with the first thing to come to hand—one of his dog collars. Perhaps he could not bear to look at her face but there was no explanation of why her body had been so well preserved and, as many villains are also liars, would anyone believe him anyway? The privet, poplar, elder, damson, goosefoot, nettle, spores of *Sphagnum* moss, and alien garden plants I did not even have to identify whispered to us of where she had spent her last lonely hours, and helped in the conviction of the man who killed her.

CHAPTER 8

Beauty in Death

For some years, I lived with my grandmother and her elderly cousins in their grand home in Rhyl, in North Wales. There were large eaves formed by the roof overhang, and these were a haven for wildlife. I was used to the chittering and fluttering sounds of bats roosting under them and, on one warm, balmy night, I had my first close encounter with one. It was a particularly hot summer, and the bedroom window was wide open, with the curtains drawn partly back. I was startled awake by my grandmother flapping around the bedroom with a rolled-up magazine in her hand, seemingly hitting out at nothing. I sat up, rubbed my eyes, and was startled to see that she was chasing a bat. The poor little thing had unwittingly come through the window but was now soaring in blind arcs around the bedroom, skimming at incredible speed between and around obstacles, the most potent one being the weapon in my grandmother's hand. In her diminutive grasp, the rolled-up paper was as strong as any wooden truncheon. She took a fatal swing and the bat dropped out of the air above my bed. Stunned or dead, I did not know. In a flash, she picked it up and tossed it out of the window.

Relieved, she came back to the warmth of the huge double bed we shared and was soon asleep. As I lay awake beside her, I was upset. Why did she kill anything she didn't like? Nowadays I know that, being born and bred in Australia—where anything and everything might kill you—she could never take the risk of leaving something as alien as a bat flying around our bedroom. She was probably convinced it would suck our blood as we slept.

Immediately after breakfast, and as quickly as I could, I ran around the house to just below our bedroom window. There lay the bat, still—dead. I knelt down to touch it, slightly fearful for I had never been close to one before. Its fur was a wonder of softness. I lifted its wing and its little clawed foot hooked onto the end of my finger. I was fascinated to learn, just by looking, that the bat's wing was actually a "hand." Instead of feathers, the wings were of very fine, thin, dark leather stretched between its long, slender fingers. This animal was very beautiful, and its death made me cry. But I cradled it inside, wrapped my clean socks around it, and hid it in the drawer by the side of the bed. I had spent the past weeks studiously teaching myself to knit, so I ran to my wool bag, unpicked a knitted square that had taken me an age to make, and carefully wound the pale blue wool around and around the little corpse until it was cocooned like a mummy. There followed a solitary cortege to a quiet part of the *Fuchsia* hedge, where I buried the body under a mass of scarlet, pendulous blooms, digging out the soft soil with a spoon I had hidden in my pocket. I have never forgotten that wanton waste of a lovely little animal's life and, for the first time, looked upon my grandmother, Vera May, as being less than perfect.

What happened to that bat as it lay in its woolen shroud, three inches under the earth, is similar to what will happen to us all. The romance of death, so saturated in the art and poetry of bygone days, is so false. One day, you and I will both be the same as the bat: just lifeless flesh, blood, and bone, and the wonderfully intricate workings of the body will just stop—dead.

I was born in a time and place where we did not question the existence of God or the worth of religion. As a girl, I was a regular chapel-goer and did not question that Jesus Christ had died for our sins. No one else seemed to doubt it either. If you were good, you went to heaven and if you were bad, you went to hell. But the multiplicity of colors in life experiences fueled my rejection of such simple, black-and-white ideas. I slowly began to realize that life is difficult, complicated, and certainly unfair, and I could see no logical reason for an afterlife where I might exist in perpetuity. Logically, the only afterlife is the handing on of one's genes to offspring; but, one can still attain immortality through leaving one's writing, art, or music. I came to believe that there is no soul eternal and, although my conversion was gradual and obscure, I know I will end this life as a committed, perhaps even fundamentalist, atheist. I firmly believe that we exist because of chemistry and physics, and our physical being will be recycled in the same way as it always has been.

Your body is your own for only a short time; the elements from which it is made are only borrowed from the outside world, and you must give them back eventually. The entity that you recognize as *you* is a collective of ecosystems that many different types of microorganisms call home. And although you might die—when your brain and circulatory systems have irrecoverably stopped working—the communities of bacteria and fungi, and even mites in your pores and worms in your gut (if you have any), will live on for some time.

Soon after your blood ceases to flow, your body will cool until it reaches the ambient temperature of the place where you died. These environmental conditions will have a significant bearing on what is to come. The blood in your capillaries and veins, no longer being pumped around by the beating of your heart, will settle and pool, leading to the first discoloration of your skin, a phase called livor mortis, and, after that, your muscles will inevitably stiffen, first in the face and then in the entire body, as your muscle filaments begin to bind together. This is the phase referred to as rigor mortis.

Your body will not die all at once. Starved of oxygen, your brain will stop functioning within three to seven minutes, but it can take the rest of your body a number of hours to catch up. Your skin will still be capable of being cultured in a laboratory, and coerced to grow twenty-four hours after your brain has ceased all function. It is deep inside you that the most dramatic changes are being wrought. For the millions of microorganisms that have colonized your insides, supporting the proper functioning of your body, and in particular your gut, will transform everything. Now that your heart is not beating and your lungs are not breathing, together spreading oxygen to every cell in your body, those microorganisms inside you that depend on oxygen will rapidly use up all that is left. They will fill your body with carbon dioxide and other gases, and begin to poison your body's cells. Your own cells will release enzymes that break down your tissues in a process of self-digestion, or "autolysis."

Meanwhile the anaerobic microorganisms—the ones that not only function without oxygen, but are poisoned by it—thrive and multiply. As your cells break down, the bounty for these anaerobes is immense. They proliferate wildly, the sheer mass of their growth inexorably forcing its way through your blood vessels, which provide a convenient system of tubes, ramifying throughout every tissue and organ. They use you as food, taking the energy and nutrients from your proteins, carbohydrates, and every complex compound in your body and, in doing so, produce noxious acids and gases, and many other by-products of their metabolism. Foul-smelling gases, such as hydrogen sulfide, blacken the peripheral blood vessels and contribute to the fetid smell of death. This "putrefaction" is when your body will lose the threads that hold it together, when the cohesiveness between its cells melts away, and your tissues and organs turn to mush.

Decomposition is not a constant thing. The variables that dictate its processes are many and varied and, the truth is, we are still in the dark about so much of it. My work has revealed time and time again that, just as in life, no two of us are the same; we are very different in death too. Some bodies decay slower than others. If you were on a course of antibiotics when you perished, it is likely that your decomposition will take a relatively long time: those same antibiotics you were taking to treat a chest infection have been killing and suppressing the bacteria and microorganisms not only in your chest, but in your gut too. If your gut bacteria and other residents have been cleared out by your medication, they will not be there to decompose you inside out.

The temperature of the landscape in which you died, or where you were left to rot; the moisture in the atmosphere around you; how loosely or tightly your clothes fit around you when you are laid out to decay; whether you are buried in a shallow grave of topsoil or a deeper grave of densely packed clay, or even a grave of dry, sandy earth; all of these things have a bearing on the rate at which your body disappears. Attempts are being made all the time to ascertain what speeds up or slows down decomposition but, the fact is, there are so many factors affecting the process, it is dangerous to infer too much without knowing more about the conditions that prevail while it is happening. Sometimes, even bodies interred in the very same graveyard—where the conditions seem near identical—can decay at different rates, and nobody truly knows why.

In 1998, when the local doctor Harold Shipman from Hyde, near Manchester, was arrested on suspicion of murdering his patients, nobody could have predicted the litany of horrors and secrets that were about to be revealed. Shipman was eventually imprisoned for life on 15 counts of murder—but an inquiry went on to identify 218 separate victims and predicted that, across his career, he was likely to have been responsible for more than 250 unlawful deaths. The inquiry demanded that a number of Shipman's victims be exhumed from their graves for close examination, and the postmortem pictures I saw were astounding. I remember the strong image of an embalmed gentleman, still dressed in his evening suit and bow tie; he had been in the ground for many years but was still intact and quite recognizable. One wonders whether, because of the strength of the embalming fluid that had been pumped into his veins in the undertaker's mortuary, he would ever disappear or whether he would last for thousands of years, like an Egyptian mummy. The contents of the other exhumed coffins varied; some contained very little at all, even though the victims may have been buried long after some that were still quite well preserved. It was fascinating and puzzling.

The progress and timing of human decomposition is becoming a very fashionable area of research in some universities. It is certainly popular with students who enroll in anthropology courses and have dreams about being known throughout the police forces as "the" expert to consult. Certainly, it would be useful to be able to predict how long someone had been dead just by their stage of decomposition, but the multifarious nature of the process means that it might be impossible ever to construct a universal and predictable model.

Interest in this whole area started in the 1970s when the US anthropologist, Dr. William Bass, kept being asked by the police to attend crime scenes to tell them how long a victim had been in place. He eventually said that it was incredibly hit-and-miss and it would make prediction much easier, and accurate, if he could observe real dead bodies decaying in natural surroundings. After much wrangling with local Baptists and other protestors, he was given a plot of woodland adjacent to the University of Tennessee in Knoxville, where he set up the "Facility," popularly known as "The Body Farm." This is a place where corpses are exposed to the elements and studied to see how the environment affects their decay. Enclosed by tall scavenger-proof fences and barbed wire, Bill Bass named it "Death's Acre" in his own memoirs and it has since become one of the best-known plots in the world, particularly after being publicized by the author Patricia Cornwell in her novel, also called *The Body Farm*. I was an avid reader of her early books and would have given my back teeth to visit the place but, as it happened, I did not have to sacrifice my molars; the visit was handed to me on a plate.

To aid his studies, Dr. Bass asked for corpses to be donated to science. These he would leave exposed in different conditions across the Facility and he and his students would make meticulous observations, hoping to refine what we know about the decomposition process. Imagine a situation in which police have been called to a corpse. In the real world, the body would most certainly not be laid out neatly, waiting to be retrieved. It would most likely be completely or partly buried; it may be hidden in vegetation, or perhaps submerged in water; it might be clothed or naked, bound and gagged, or even dismembered, with the parts dumped in a variety of places. At the Facility, the donated bodies are used to simulate as many different kinds of disposal as possible. They are left out in a variety of sometimes bizarre conditions and their breakdown recorded in detail. If such scenarios are repeated sufficiently often, a database can be built up of what happens under specific conditions. And, since the Facility became famous, streams of PhD students have carried out various kinds of study on the corpses themselves, the soils under, over, and around them, and the insects that colonize them. Some have set up experiments in an attempt to examine the multitude of variables that might affect the nature and the rate of body decay.

I received an opportunity to visit the Facility when, in 2005, I was asked to be the subject of a television documentary about my work. I was reluctant and, for months, refused to get involved. But the director, Maurice Melzak, the most quietly persistent man I have ever known, hit on the idea of placing part of the story in the Facility at Knoxville. He had literally pestered me, though very politely, to make a film about my work for many months. He patiently tolerated all my rejections until, one day, he asked if he could come to tea. I remember him sitting in my sunny conservatory asking, "Wouldn't you like to have a chat with Bill Bass and have a look at all the work they are doing there?" He had hit the right nerve. I wanted to go all right; I am too nosy to miss anything like that. After the slightly traumatic start to our relationship, he became one of my best friends.

I am not a good traveler and I detest the whole tedious process surrounding international airports. By the time we arrived in Knoxville, after a succession of three airplanes, endless immigration queues, shuffling, tripping over people's overlarge bags, and bawling babies, I wanted to scream. We had to run as fast as we could to make our last flight from Philadelphia to Knoxville.

We eventually arrived at the University of Tennessee and met up with Bill Bass and his colleagues. All were anthropology/archeology academics and most seemed interested in bone—the way it fractured, shattered when shot, or was preserved in various scenarios. We did the rounds of departmental offices, politely nodding to various academics as they looked up and smiled the obligatory "hello" before resuming work. One girl gave us a wonderful demonstration on how to determine the entry points of bullets, and to work out the kind of instrument matching various gaping holes in skulls. I thought her work was highly relevant to police investigations and she was a good teacher.

Bill Bass struck me as being a down-to-earth, cheery man. Born in 1928, he began his research career excavating Native American grave sites in the 1950s but has spent much of his life assisting federal and local police in identifying human remains. This is not easy: the American authorities have a great problem because of the sheer size of their country and the huge numbers of murders taking place. They will probably never know how many killings there are each year. They have several kinds and levels of policing, and it seems that many operate without support and help from the rest. The American system is very complicated to an outsider and, presumably, it is not easy to coordinate information; I have been told that there is often little cohesion or exchange between forces. Bill Bass has helped across the spectrum and the founding of the Facility in 1981 is among his crowning achievements.

On our approach to the Facility, I could feel the anticipation building up in me like a child who wakes early on Christmas morning. As we went through the heavily guarded gates, I took in a place that was cool and shaded, dominated by trees, with sparse greenery underfoot and definite pathways leading in several directions. At one edge, the ground was completely covered in kudzu vine, with green shapes formed from its relentless and suffocating growth. This plant was introduced from Southeast Asia and is now considered to be a most noxious weed in the southern states of the US. It spreads, climbs, curls, and entwines over everything in its path, so that any obstacle eventually becomes a green ghost of itself.

But that was only one of the fascinating things to greet me as I walked through those gates. Almost immediately I was in an open area where corpses had been laid out, and they were all in various stages of decay. I was intrigued to pass bodies in different postures along the pathways. Some were partially covered, others wholly covered, and some just naked. It is chastening to realize how casual one can be when confronted with corpse after corpse. I was never shocked, even though some of the sights were worse than anything you would ever see in a horror movie. The only thing that struck me forcibly was the lack of color compared to many of the dead I had seen in casework—here there was no blood; it had all disintegrated to common brown, along with the skin, hair, and nails. Only the fresh bodies seemed to be real.

The stages of a body's decomposition have been used, over the years, for determining the "postmortem interval," or the length of time that has elapsed since a person's life came to an end. The bodies bequeathed for decomposition studies are invariably of white, middle-aged, and elderly males, which, of course, means that there is an inbuilt bias in the sample of subjects being observed. Older people tend to have a greater need for medication and, of course, this might affect rates of decomposition. Black and Hispanic donors are rare, as are women, and I am not sure that the Facility has ever been donated a child. For whatever reason, black people, Hispanics, and women seem to be less enthusiastic about being exposed, colonized, and studied. To be really useful, there should be no bias in selection of corpses but, of course, you take what you can get. During my visit, I only saw one black corpse laid out in the woodland, and his story was rather sad. His family did not want to pay for his funeral and thought that donating him to Bill Bass was a useful solution. In another case, the individual had been so horrible in life that his family wanted to "punish" him in death. I remember being grimly amused when told the details of his story but, in reality, it was sad for everyone concerned.

As we walked quietly along, one of the senior female academics said, "Oh, avoid the vine—it's full of copperheads." I had heard the name before and assumed they were butterflies. Wrong: the copperhead is one of the pit vipers and, although its bite rarely causes death in adults, the effects are very nasty indeed. One amazing fact I learned about these snakes is that, in the absence of a mate, a female can produce live young without sperm. Her egg cell will divide twice to make four cells, and two of these join to make an embryo. So virgin birth exists even in the vertebrates as well as invertebrates! I might never have come across that if I had not visited Knoxville, but I was to learn even more about the menagerie of nasties in Death's Acre.

I had noticed that all the people from the university carried long sticks. I later found out that these were for knocking down webs of the brown recluse spider. These spiders are venomous and terrifying to most Americans, and their bites have serious consequences. Textbooks say that they are not as dangerous as people think. Tell that to someone who has to work surrounded by them. I have three fears which, I am sure, I share with many others—heights, spiders, and snakes, and two of the three were all around me. I was not a bit worried about the bloated bellies bursting with maggots; or the eyes and nostrils being so stuffed with fly eggs that they looked as though they were filled with cotton wool; or the smells; or empty eye sockets, gaping mandibles, and hair slipping away from the scalps. But I was certainly anxious that I might have a close encounter with the wildlife.

Neither the spiders nor the snakes got me but, ironically, an innocent-looking plant certainly did. I was wearing thin, cotton, cropped trousers and, back at the hotel, I noticed red spots on my legs. They were as itchy as flea bites but sore too, and I could not leave them alone. I thought I might have picked up jiggers, a red mite whose larval stage digests holes into the skin and sucks up the resulting "soup." But no, it was confirmed by a local to be poison ivy (*Toxicodendron radicans*). That night was one of the most miserable of my life and, at about 3:00 a.m., I tore off my night-clothes to reveal running sores on my legs—I had been ripping at them with my nails all night. Instinctively, I ran a hot bath and scrubbed the agonizing, burning skin frantically with a nail brush and shower gel, then put my legs under the cold tap. After that drastic treatment, and dabbing with calamine lotion, my legs felt much better almost immediately. I later learned that the culprit in the plant is an oil (urushiol) that can stick to skin and clothing and, as soon as the oil is removed, the skin can heal. There is a saying that probably every American child knows: "If the leaves are three, let it be." Poison ivy has three leaves on the end of its stems. I will never forget it, and I have the scars to prove it.

We were back at the Facility the next day to film a final-year PhD student and chat about her research over one of her corpses. The one that interested us that morning had been fresh and very attractive to the kinds of flies that come early in the succession of scavengers and feeders of flesh. The student was studying a suc-cession of fly communities in an attempt to catalog the sequence exploiting the rotting meat. It usually starts with bluebottles such as *Calliphora vomitoria*, and greenbottles such as *Lucilia sericata*. Of course, different species are involved in different parts of the world, although in a great many places, bluebottles and green-bottles seem to be very common pioneers on a fresh corpse. They can find a body within minutes of it being laid out and the females immediately start laying eggs in every exposed orifice. They are programmed to go into dark places and, sometimes, even make it right up into the higher nasal passages.

All scientific knowledge is incremental but, once the docu-mentary was complete, I came away with a clear sense that, as pioneering and exciting as the Knoxville Body Farm was, it could only ever be one component of our quest to understand what happens when a body decays. One site in one climatic regime, on one kind of soil, in one kind of woodland, was never going to result in an all-embracing model for decomposition of the human body. That is why, since the original Facility started up in Tennessee, six others have been established in the United States, one in North Carolina, Illinois, Colorado, and Florida, and two in Texas. There is also one in Australia, just outside Sydney. Scientists in the UK are trying to get one established too, but there are always people resistant to such ventures. At least the existing ones now offer different ecological conditions—different temperatures, humidity, soils, microbes, and scavengers.

It is going to be a very long time before any truly useful predictive models can be constructed. But body farms have been a helpful tool in advancing some of our understanding of human decomposition, even though we must always add caveats to the results. For years and years, scientists and their students in the UK, and other parts of Europe, have been using pigs as analogs for human corpses in body-decay studies. The decomposition of a pig is interesting to watch and, again, there are always differences between individual animals, but I have always doubted the validity of using a pig's corpse as a substitute for a human one. They appear to be similar to us in many ways and, apparently, we even taste like them when cooked, but one of the barriers to decomposition processes is skin . . . and pig skin is much thicker and tougher than our own. What is more, pigs generally have more subcutaneous fat.

These differences affect the way that scavengers and microbial populations can influence the early stages in the decay process. Like others, I have buried pigs in various places in an attempt to explain certain events; but I have only done this in relation to specific cases when trying to re-create a sequence of known events, in a known soil and at known temperatures. This work was enlightening and definitely worthwhile for one particular case. I was able to prove that a murder victim buried in a fairly acidic clayey woodland soil on Christmas Eve would, in most cases, lie undiscovered by dogs and foxes until April. By the usual technique of identifying the largest (oldest) maggots on the corpse, the original entomologist on the case had claimed that the victim had died in February; but police intelligence told investigators otherwise. I was asked if I could test the results.

I used three pigs that had been freshly killed for veterinary research, provided by my ebullient Irish friend Helen O'Hare, who was reading veterinary science at Cambridge, and I will never forget that cold, dark Christmas Eve when Helen and I, helped by a group of enthusiastic police officers, buried the still-warm pigs at the crime scene. The findings of this experiment suggested that, indeed, the victim could have been buried on Christmas Eve. I presented a theory involving soil effects and published the findings with the King's College entomologist as coauthor. That paper seems to have become a classic and is frequently cited. There are many published studies of buried pigs that, in my opinion, have little hope of being truly useful. Experiments need to be carried out within the context of specific investigations as this one was. We managed to glean a great deal from that simple experiment, but it involved my visiting the buried pigs every week for months, and taking soil and air temperatures continually with a thermometer linked up to a computer.

It would certainly be useful to be able to construct a robust predictive model for the way human corpses break down. But there are so many factors influencing the phenomenon that such a useful model is probably a long way off. There are just too many imponderables and broad rules become impossible to apply. I feel rather uncomfortable, therefore, when I read a description of a corpse as being in the "putrefaction stage" or in the "bloated stage." Some corpses bloat while they are in the putrefaction stage, but some never bloat at all. There have been many clever techniques applied to the estimation of postmortem interval, and most of these involve observing sequences of changes in body tissues and fluids. One technique involves the chemical analysis of the vitreous humor in the eye, others examine assemblages of amino acids, fatty acids, volatile organic compounds, or ammonia, uric acid, lactate, and many other compounds generated after certain time intervals. A good deal of information has certainly been derived from the body farms but what I find most interesting is the utter variability of it all. Whatever else we might find out about the processes our bodies go through after we die, death itself is the only absolute biological certainty.

Some years after my visit to Knoxville, the director Maurice Melzak became seriously ill—and, for a while, between us, David and I nursed him as he faded. The specialist cancer hospital was close to our home and we took him there, bringing him home most nights, wrapping him in warm domesticity and comforted by our black cat, Maudie. He was only sixty-three when he succumbed to that most pernicious disease, and I miss him even now. He was an inventive, good man, in love with the natural world. I grew to love him too, and his rather eccentric manner.

My trip to the Facility had given me a friend, but it also gave me a greater insight into how little we know, and how much we have yet to discover about what happens to our bodies after we die. As mentioned, the possibilities, from putrefaction to mummification, are vast; and, as if to prove it, one particular story sticks in the mind, possibly because it happened in Wales, so close to where I was born.

There was a family whose lives centered around the pub that they ran, a pub that had been in their family for a very long time, and a place that was the center of everything in that small Welsh village: a meeting point, a social spot, a place of celebrations and commiserations. It sat alongside the family home and an outhouse that had a mezzanine floor, but was mostly used for storing crates and barrels on the ground floor. For years the father and mother of this particular family had run it together and, eventually, they shared the duties with their son and daughter-in-law, a welcome addition to the business. Then the wife went missing and everything changed.

People go missing all the time and, in any given year in the UK, there might be a quarter of a million missing persons reports filed with the police. Almost all of these people will be found again, most resurfacing in their own lives without the drama of police and urgent searches. But there will always be cases of people who walk out of their lives, or vanish under other circumstances, never to be seen again.

When his wife disappeared, the pub owner called the police and they went through the motions, searching high and low. Friends and family, even distant relations, were contacted in the vain hope that she might have been seen. But all across Wales, and out into the broader world, the story was the same: no one had seen her. She had truly vanished without a trace. The husband said that he was convinced she had gone abroad because they had not been "getting on," and that he just had to accept it. Years went by and there was never any sign of her. She seemed to have just disappeared. But there were those in the village who never really believed him, and there was always a cloak of suspicion wrapped around this man. Many were convinced he had "done away" with her, but there was no proof; searches came to an end, interest eventually faded, and life went on as before. The memory of the missing landlady also faded and people, including the pub owner and his son, just carried on with their busy lives.

Twenty years after his wife's disappearance, the pub owner died himself, and the son wanted a new start. He and his wife decided to refurbish the old pub, make it more popular and put some zing into the business. The barrel store with the disused mezzanine loft had always been a bugbear of his. He thought it was a waste of space and that it could be put to better use. The only way up into the loft was via an old ladder that had always been stacked behind the barrels. One weekend, after closing time, he decided to have a look at the state of the upper part of the building. He climbed the ladder but, as his eyes came level with the floorboards of the upper story, what he saw shocked him so much that he nearly fell off. His mother had not run away after all.

As it happened, when the body was discovered, I was in Gwent visiting my mother and her husband—the man she had married many years after she and my father had divorced. After two days of avoiding quarreling with her, I was glad when the police rang. It was not too far to the pub and I could easily get there and back in a matter of hours.

The pub owner's son had found his mother as a mummified corpse, only a few feet from where the ladder was touching the floorboards of the loft. As I took my turn to climb the ladder, a grotesque picture greeted me. There was a grinning skull, wrapped in what looked like an old rug, and it seemed to be staring right at me. The forensic pathologist followed me up into a room that was shockingly neglected. Ivy was growing vigorously through holes in the roof and creeping along stacks of wood. Everything was overwhelmingly dirty and dusty, just like an exaggerated staged scene in a horror film.

On the day she disappeared, she must have thought that life was not worth living and had crept up into the loft, lain down and swallowed as many pills as she could. The empty bottle was just beside her where she had dropped it as her hands went limp. Carefully placed at her side there was an old-fashioned, wide-necked milk bottle, still cloudy on the inside. Bizarrely, a pair of dentures lay deep in grime about a meter from her head. The pathologist suggested that, after taking so many pills and liquid, she had vomited violently before she died and her dentures must have shot out of her mouth. She was not wrapped in a rug—it turned out to be her woolen car coat and, pathetically, her head lay on a rolled-up sack. She had obviously not cared about comfort; she had just wanted to die and get away from all the strife in her troubled life. One could only wonder at the torment she suffered when alive. Like a wounded animal, she had crept away to die as anonymously as possible.

The loft was incredibly dry considering there was a hole in the roof, and there was a continuous breeze wafting through. This had acted to dry her body quite quickly so that, instead of her belly becoming bloated and her skin and hair becoming wet and detached, she had slowly dried out and mummified. Now she was a desiccated, leathery version of herself. A good deal of information has certainly been derived from the body farms, but bodies seem to be predictable only in their unpredictability. After death, this woman's decomposition had not followed the pathway I had seen in so many victims dumped in woodlands, ditches, or even left in armchairs. Her gut flora would have contributed to the decomposition of her internal tissues but, even inside, we later discovered the remains of her organs. Flies must have found her because there were some puparia around the body—but they had all hatched into adult flies and were long gone. Fairly quickly, she must have become too dry to tempt further egg-laying.

The next morning, I arrived in the hospital mortuary in Cardiff for the mummified woman's postmortem. Every examination yields different information about the life and death of the dead person and, for a change, it was a relief to work on a corpse that did not stink. But the story of the suicide in the attic has a grisly coda. When she was opened up, the incredulous pathologist drew out what he believed to be a string of beads from inside the rib cage. I was almost as disbelieving at what I saw, even though I could identify it immediately. This was the desiccated gut of a rat still with its fecal pellets; they were regularly spaced like a string of beads, and this made it almost comical. I poked around a bit farther up into the cadaver's thorax and picked out a beautiful little rat skull.

We can only theorize as to what had happened. It is possible that, by eating the flesh which, by this time, was saturated with barbiturates, the rodent had become poisoned too, and had died inside its cocoon of plenty. What was strange—really strange—was that we could not find any other rat bones. Where had the rest of the skeleton gone? Had some other scavenger, possibly a cat, found the rat in the cavity and made off with its booty, leaving just the head behind? It was a case of Russian dolls, one drama inside another, and another.

Putrefaction can be delayed by any number of factors, natural and artificial. Poisons like arsenic and strychnine can arrest the decay, as can antibiotics and other medication, but the ambient environment at the time of death can also preserve soft tissue. Natural mummies like this might, given the correct conditions, last for thousands of years. In 1991, two German tourists were hiking along the Austrian-Italian border, through the Ötztal Alps, when they came across a mummified corpse preserved so perfectly that they presumed it belonged to a recently deceased mountaineer. The bottom of its body was frozen in ice, but above the torso, the man remained much the same as he was on the day that he passed on. When the mountaineers raised the alarm and the local gendarme arrived to recover the body, it was taken to a medical examiner in nearby Innsbruck—and it was here that the body was finally dated as being at least four thousand years old.

Ötzi, as the man came to be known, had lain here on the mountainside since the Neolithic period. He was a Stone Age man. The low temperatures and high winds had resulted in preservation of not only his tissues but also his clothes and footwear, his bow and arrows, the food in his pouch, and the fungus he had been carrying as tinder. Analysis of his stomach revealed his last meal, and the arrow, that has recently been shown to have caused his death, was still stuck in his shoulder blade. We only decompose if the environments inside and outside us favor the growth and activity of microorganisms. If they are inhibited, a mummy can form. Many such mummies have been found in the steppes of

Asia and in the mountains of South America, where conditions are too extreme to allow microbes to be active.

If the stages of decomposition cannot be relied upon to give us accurate information about the timing of a person's murder, then we must use other clues. And, in this, forensic ecologists have another important weapon in their arsenal. This weapon is a whole kingdom of organisms that evolved on Earth before the fish in the sea, insects in the air, or animals and plants on land. They are everywhere—in the soil, inside and outside plants and animals, colonizing our countryside, our gardens, our homes—as well as in and on our own bodies. They have often proved to be the source of the most crucial information. These are the fungi.

CHAPTER 9

Friends and Foes

I was educated in the botany department of King's College London, and mycology (the study of fungi) and bacteriology were two areas of study at which I excelled and most enjoyed. Traditionally, the fungi include mildews, molds, blights, yeasts, lichens, rusts, smuts, slime molds, and of course mushrooms. Until fairly recently, they were studied by botanists because they were thought to be plants, and it was the Swedish botanist, Carl von Linné, in his two-volume *Species Plantarum*, published in 1753, who promulgated for this monumental biological mistake. As he worked out the various groups of known living things, he put them into named categories. He decided that fungi were plants—and plants they stayed for two hundred years, well into the 1960s. It was only then that we made the conceptual leap: fungi are now placed in their own kingdom,

Fungi, while some blights and slime molds have been moved out to the algae and protozoa, respectively.

Fungi diverged from the rest of life about 1.5 billion years ago and recent molecular studies show that they are more closely related to animals than plants. They certainly feed in the same way. Like animals, they can only survive on food that has already been made, mostly by plants or organisms that have, ultimately, fed on plant material. They are also like some animals in that digestion of their food occurs outside their bodies. The spider traps its fly, pours digestive enzymes onto it, and these dissolve the unfortunate prey into a mush. The spider then sucks up the liquid and discards the insect husk. Fungi do a similar thing and, indeed, some that live in the soil produce lassoes which catch tiny nematode worms. The fungus then grows throughout the worm and passes out enzymes to digest the prey's tissues; these are broken down to molecules and are absorbed by the fungus into its own body. When a fungal spore germinates, it will develop a thin thread (a hypha), and this branches repeatedly to form a radiating, interconnecting mass of thin threads known as mycelium. A fungus has the advantage that it can push into and through its food, as well as grow all over its surface. They are certainly not plants; plants use the magic molecule, chlorophyll to trap the energy of sunlight and use it to convert carbon dioxide and water into sugar. Like all animals, in order to grow and reproduce, fungi need to feed on the living or dead tissues of other organisms.

Fungi are very ancient indeed, but they do not fossilize well. Even so, there is some evidence of fungal-like organisms as far back as 2,400 million years ago. About 542 million years ago, and long before plants managed it, they had colonized land from the sea. And, in the Silurian period, starting about 444 million years ago, they were already highly diverse, occupying many ecological niches. To put them in context, fungi were well established about a billion years before dinosaurs first walked the earth.

Taxonomists agree: we know most of the plants and animals—except for nematode worms and some insect groups such as beetles, whose number of species just keeps on growing—but the situation with fungi is more startling. With new molecular studies, it has been discovered that each fungal species that has been described probably consists of five or more species. Recently, one species of *Aspergillus* (a genus that includes serious human pathogens) was analyzed and found to be forty-seven kinds of organism, each with different potential. The scale and diversity of the fungal kingdom is vast, and as yet we know barely 5 percent of the species sharing the planet with us.

Plants build up and fungi break down. Fungi are the principal agents of the degradation of plant material. Indeed, they are the only ones that can break down the lignin, the complicated polymer that makes wood hard, but they also play some role in the decomposition of all dead things. Where dead things lie, the fungus feasts. If it did not, and there were no breakdown, all the chemical building blocks of life would remain locked up forever in the dead bodies of plants and animals, including those of humans. If this were to happen, nothing could ever be born and life would come to a standstill. Decomposition is utterly essential if we are to have any life at all. Living things are constantly being recycled and, eventually, that will include you and me.

Until recently, forensic scientists believed that the only way fungi could be used in their investigations was in cases of poisoning, or the illicit use of psychotropic (hallucinogenic) species. But what fungi present is actually a much richer cache of information. The way they grow, the rate they grow, the patterns their growths make—all of these can be recorded and interpreted to help the canny observer put a certain person in a certain place at a certain time, to estimate the length of time that has elapsed since a victim drew his last breath, to ascertain the actual cause of death. Like pollen, fungi leave their testimony wherever they grow.

Fungi can be microscopically small and improbably vast. The hyphae germinate from microscopic spores and these threads aggregate to form the spreading, interconnecting, and outwardly ramifying fungal mass of mycelium. These threads only stop when they meet a barrier or run out of food. If there is nothing to prevent it, and it can scavenge food, the mycelium can spread for miles and miles for very many years. There have been several reported huge colonies of species of *Armillaria* (honey fungus) in North America, but the record holder for the largest is a giant *Armillaria ostoyae*, discovered in 1998 in the Malheur National Forest in Oregon. Based on its current growth rate it is estimated to be about 2,400 years old and could be as much as 8,650 years old. It covers nearly four square miles and, as it radiated out from the point where a microscopic spore germinated, it killed trees and then fed on them so that it was never short of food. Besides the dead trees, its presence is revealed by honey-colored mushrooms— pretty masses emerging from the base of the tree trunk.

This is one of the world's vast fungi. But the black mold, *Cladosporium*, growing between the tiles of a poorly ventilated bathroom is a fungus as well. So are the green and white patches that appear on a loaf of bread you have left uneaten for too long, or the variously colored greens that appear on an orange left uneaten at the bottom of the fruit bowl. The yeasts that bakers add to flour and water to make bread, or that brewers use to make beer—these too are fungi. Without fungi, we would have few antibiotics, no fizzy lemonade, biological soap powders, tea, coffee, trees, flowers, most foodstuffs in the larder, and many essentials in our modern lives. Even many animals that we eat would not thrive without the fungi in their guts, and the grass would not grow for them to eat in the first place. We are surrounded and penetrated by fungi and we cannot survive without them. We feed them and they feed us.

Wherever there is digestible food to be found, fungi will follow. A single teaspoon full of soil from the top of the earth's crust can contain more than 100,000 viable spores and little bits of fungus, each one capable of making a colony. Think of this the next time you look out of the window and really consider the world around you. Our own bodies, and most mammals, are covered in yeasts, such as *Malassezia*, which does not usually do any harm, although one species is the most common cause of dandruff. And, of course, many people are aware of the thrush fungus, *Candida*, which can cause irritation but, on some rare occasions, death, should it get into blood and internal organs. Most people get some sort of fungal infection in their life. If you've ever had athlete's foot, you will have been colonized by a species of *Trichophyton*, or possibly *Epidermophyton floccosum*. These are common soil organisms, so it is sensible to protect the feet when working in, or walking on, bare soil because these fungi can make the foot (usually between the toes) very sore and scaly. When *Trichophyton* causes circular patches of red, sore, scaly skin anywhere else on the body, even the scalp, it is called "ringworm"—although, of course, no worm is involved. Fungi are notoriously difficult to get rid of because most of the medication that will kill them will kill you too. Bacterial infections are fairly easy to dispel because bacteria are so different from us in every way, but fungi are more closely related to us, and we share some of their sensitivity to certain toxins.

If the living human body, with all of its defenses and immune system in good order, can provide such a fertile habitat for fungi, what then of the body after it is dead? To many species of fungus, a human corpse is a vast source of nutrients waiting to be digested. And in one particular case that has always lodged in my mind, it was the way in which fungi had feasted and spread upon a murdered man's carpet and sofa that eventually contributed to confirming his killer.

Picture a block of flats in a dismal, wintry Dundee. Picture a front door forced open by police after someone has reported their friend missing. See a man lying spread-eagle, his face pressed into the carpet, and the multiple stab wounds—no mystery at all around the cause of death.

Now picture the blood and other body fluids that have splashed on the furniture as they spurted out of the stab wounds, and then oozed out of him, soaking into the carpet. Windows are closed and the flat is incredibly hot because the central heating has been blasting out for some time. The blood spatter is highlighted by gray, white, green, and brown growths of fungi whose spores, lying dormant in the furnishings, have been kissed awake by the sudden provision of all this food. Until now, the flat has remained closed; the corpse has been protected from scavengers, and the flies which might otherwise have come and laid eggs in its orifices. Fungi have spread wherever there was food—but, now that the blood and bodily fluids have dried up, or been entirely used by the fungus as it spread, its growth has come to a halt. The colonies are like an atlas on the carpet, marking the shorelines and boundaries where the dead man's blood has reached.

This was the scene that faced me when, in 2009, my husband David and I caught the night flight to Dundee and arrived at the dead man's flat. One of the police officers had been told that fungi could be used to estimate the timing of events and, bright chap that he was, recommended to the Senior Investigating Officer to get us to the scene.

We often get weird and wonderful puzzles to solve, and the key to deciphering what had happened here was the extent the fungus had grown in the blood that had spilled from the man's wounds. The first thing to do was decide on a representative sample area, take photographs, and draw diagrams of the colonies within it. I then cut out representative pieces of the sofa cushions and the carpet, and put them into sterile plastic containers. By telephone, we had already asked for humidity and temperature to be monitored, and a substantial record for the flat was already available. It confirmed that the average temperature had been about 26°C and the relative humidity about 34 percent. Lovely and warm for fungal growth, but definitely too dry—most fungi need a relative humidity of around 95 percent to grow at all. You will know this from your own home. If you get a leak from the roof, the wallpaper will soon go black and green with fungal growth. Where you have mold growth, you have damp. What this meant, here in the flat, was that until the dead man's blood had splashed out over the carpet, there had not been enough moisture for the spores, dormant in the carpet and furniture, even to germinate. Then a bonanza of food and water had become available.

We asked the police if we could use facilities at their forensic laboratory and, although they did not have anywhere dedicated to microbiology, they did have a good laminar flow cabinet which would protect my preparations from stray spores in the air. As usual, we had all our small pieces of equipment with us, and I set up a makeshift inoculation facility in the cabinet. All I needed was a Bunsen burner for sterilizing and Petri dishes with some basic medium. I carefully made cultures of a number of colonies onto plates of medium.

For a long time, David had been the director of the International Mycological Institute at Kew and, for most of his working life, he had had technicians to carry out the drudgery of the technical work—sampling, subsampling, culturing. I, on the other hand, had never had the luxury of a minion to carry out work on my behalf, so I have always retained my practical skills as a microbiologist—and, after photographing and drawing the patterns of fungal growth on the various fabrics, I defined representative areas for sampling and measured all the most clearly defined colonies that were in my drawings. I then cut out the samples of fabric, cultured each colony onto agar plates and put the samples into sterile containers. We then flew back south with our moldy bits of carpet, cushion, and precious cultures, and set about an experiment to understand what the fungal growths represented in that apartment.

Our first step was to incubate the moldy fabrics at the same temperature as the flat. When, four days later, the fungal colonies looked exactly the same as they did when we originally cut them out of the cushion and carpet, it was obvious that their growth had been halted by something, and that something was probably a lack of available water.

To test this point, we wetted all the fabrics with bovine blood and incubated them overnight. By the following morning, fungal growth had been explosive. Each fungal colony had spread out, fighting for space, overgrowing each other wherever they could, and there were no gaps at all. Now we had a working theory. The warmth of the flat had dried out the body fluids and, as soon as it became dry, the fungi just could not spread any farther. By measuring the size of colonies growing on the carpets we could establish the minimum amount of time the body had lain there if we knew the rates of growth of the various fungi.

This meant taking each pure culture, prepared in Dundee, and subculturing it in the middle of a new Petri dish of growth medium. The fungus would then grow outward from the inoculum to form a circle. We then watched, measured, and timed the spread of the three isolates, which David had identified as *Mucor plumbeus*, *Penicillium brevicompactum*, and *Penicillium citrinum*. The incubator was set at 26°C and, after two days, the *Mucor*, being a fungal "weed," had virtually covered the whole plate as we expected. It took five days for the two *Penicillium* species to reach the same diameter as they had reached at the crime scene. This meant that the blood had been spattered on the cushion, and seeped onto the carpet, a minimum of five days before the body was found. What really amazed us was that the offender, a "friend" of the victim, later admitted the crime and confessed that it had been committed five days before the alarm was raised. If the windows had been left open, flies might have got in and, inevitably, the police would have asked an entomologist to estimate the time of death. Yet again, we had shown our usefulness when the tried and tested methods just do not work.

Fungi can also provide excellent primary trace evidence, or even provide corroborative support to botanical or palynological evidence. They are useful because they can grow wherever there is even a tiny splash of food—on glass, paper, wood, leather, and even plastic. Fungi have adopted many different kinds of lifestyle. They may simply feed on dead organic matter, and this is why the leaves in an orchard disappear by the spring, or they may invade and parasitize a host, even killing it so that they can feed on the dead body.

Some fungi form intimate and mutually beneficial relationships with plants. The plant feeds them sugar and the fungus gives the plant phosphate, water, and other nutrients. As the fungal mycelium spreads out into the soil, it effectively extends the root system of the plant. I find it even more wonderful that one such fungus can form the same relationship with several plants at the same time, and food can pass throughout the system. This means that plants can be linked, and one that is struggling might be fed by another plant via the fungus. Thus if a tree is doing very well at the edge of a woodland and one is growing poorly inside, the outer one can pass food to the starving one. When you think that each plant can form relationships with many species of such mutualistic fungi, we can think of vegetation as being all linked up rather than behaving like single individuals. Today, ecologists increasingly talk of the "wood-wide web." The world is so complicated and so incredibly wonderful!

Not all fungi are benign, though, and many are desperate killers. They do not live in harmony with their hosts but invade, kill, feed, and move on, either by spores flying into the air and landing on leaves and stems or by creeping through the soil to attack roots. Animals also benefit and suffer from fungal infections. Those that live in the animal's gut actively digest indigestible foods down to simple molecules that can be used by the host. Cows, sheep, and goats chew the cud to make indigestible food like grass, hay, and leaves more accessible to the huge volume of microorganisms in their guts. They would starve without them. The wild rabbit even has to eat its own feces because the place in its gut where microbes do the digesting comes after the place where absorption occurs. What a freak of nature that is.

Another startlingly robust symbiosis of fungi and plants—and, sometimes, a fungus, plant, and various species of bacteria—is the lichen. Lichens are the scurfy, leafy, or even bushy growths, invariably in shades of gray, green, and black, found on rocks, buildings, walls, tree trunks, leaves, and even on the ground. One looks just like trodden-in chewing gum and, where it grows, a pavement can look really polluted. The lichen partnerships are incredibly ancient and go back hundreds of millions of years. They are exquisitely balanced associations and they can tolerate the extremes of physical and chemical conditions on this planet. They are found in Antarctica and arid deserts but are most diverse in hot, steamy jungles and temperate forests.

Lichens are the ultimate survivors and have survived space rockets that have crashed to the ground in flames, and a specimen of *Xanthoria* was attached to the outside of the International Space Station for eighteen months and survived cosmic and ultraviolet radiation, and the vacuum of space itself.

Lichens are ancient and durable, and each one is certainly not an individual organism (any more than you are); it is a microcosmic community of a fungus, one or more algae, and bacteria. Like other fungi, they can be used to give a good idea of how much time has elapsed. Even geologists trying to estimate the growth and decline of glaciers have used them to estimate changes through time. And so it was when, a few years ago, David and I were at a biodeterioration conference in Manchester and the call came through that we were wanted urgently. On a quiet stretch of road, bordered by woodland and 190 miles away, a truck driver—stopping on the side of the road to relieve himself—had discovered a suspicious bag that he thought might have been a dismembered body part. A murder case, in which the victim had been dismembered, was dominating the news at the time and the public had been asked to look out for anything suspicious.

We were told a bare outline of the case. A man and woman envied the financial status of an associate and decided to kill him. They took over ownership of his house and car, and their naïvety in thinking that they might get away with this is overwhelming. The hapless victim had been expertly butchered and his body parts distributed over a wide area. His skull was found in Leicestershire, his torso was found in a suitcase in a Hertfordshire stream, and the arms and legs were variously found on verges, woodlands, and fields. I had worked on several of the discovered body parts, including the skull, but the power of botany and mycology for one leg made an impression on many.

David and I left the conference in a hurry and drove, as fast as speed limits would allow, down through the Midlands, past Birmingham and all its traffic congestion, to the place where the leg had been found. Cordons had already been set up and, by the time we arrived, the crime scene investigators had already removed the leg to a mortuary; we missed it by twenty minutes. To this day I will never understand why and, pretty fed up at this, and with so little we could do in the fading evening light, David and I decamped to a local bed and breakfast, utterly exhausted after the conference and long drive. The next morning, bright and sharp, we arrived at the site where the leg had been deposited. I started by doing a vegetation survey, assessing possible offender pathways, and taking soil samples for possible comparison with the footwear and clothing of one or more offenders whenever they might be apprehended. When the truck driver had found the leg, it had been wrapped in blue plastic. He claimed not to have touched anything, saying that he just telephoned the police as soon as he realized there was something suspicious—but, when I looked at the ground, it was obvious that he had had a good poke around the parcel and had certainly moved it about a meter from its original position. How did I know? Well, the original position of the leg was quite clear to me; rodents had nibbled at the blue plastic and earthworms had already started to bury it in their casts.

There were several small herbaceous plants that had obviously been bent over by a weight, but they were still green and had started their recovery growth. The chewed-up bits of plastic and the little herbs were about a meter away from the putative position of the parcel; crime scene investigators had carefully marked the spot they had found it. I already knew that the truck driver had bent the truth slightly and that his curiosity had resulted in displacement of the parcel. I was doubtful that it had been there for more than a few days because of the state of the plants, and burial of the comminuted plastic was fairly minimal compared to some I had witnessed in other cases. The police investigation was getting grislier by the day. This was not the only body part turning up in a remote location, and David and I had already been to a site to examine the head in Leicestershire, and an arm and another leg at different locations in Hertfordshire. The police already had their suspects and, after interview, had come to the conclusion that the murder had been committed two weeks before the last leg had been found. The assumption was, therefore, that all the body parts had been dumped at the same time, about two weeks previously.

I was kneeling down, looking carefully at the plants that had been affected by the bag, when David poked me in the back and pointed to a twig that was lying on the stem of one of the plants at the original site of the bag. Following his finger, I saw some extensive colonies of *Xanthoria parietina* on the twig. This lichen is bright yellow with orange spore-producing bodies when in full light—but gray and orange when in the shade, as on the underside of a branch. It is incredibly common in the south of England, especially near roads, where its growth is stimulated by nitrogen pollution from vehicles. But that was not what interested us here. The colonies on the twigs were indeed yellow, and David knew from field observations that they became greenish if covered, although he did not know how long that took. We felt this might be important, but in that moment we had little idea how useful it would become. We took the twig from the ground, and also some others on nearby branches, and the attendant police officer logged them out of the scene.

There was nothing more we could do until we brought the twigs home to our garden. We live next to one of the biggest oak trees in Surrey and its leaves are a nuisance toward the end of each year, but now they were useful. We allow the garden to run wild in our little orchard so that the foxes and badgers have somewhere to sun themselves, or to roll and play, as we have seen them do many times. It is also useful for setting up simple experiments to test ideas, and we decided to do this here. David knew that, when in full sun, the lichen was yellow and orange but, when covered, for example, on a branch that had been turned over, it turned green and eventually died. Determining the speed at which this particular lichen might turn green when deprived of sunlight might help us to work out how long the dismembered leg had lain in the woodland.

We set up a little cage from chicken wire to prevent any accidental disturbance, and we put a thick layer of dead oak leaves on the ground to simulate the leaf-strewn woodland floor where the leg had been dumped. Next, we took three pieces of twig, each bearing a good sample of the lichen, and placed them onto the leaves. The first twig we left completely exposed to the light, while covering the second and third twigs with blue plastic bags full of sand to simulate the weight of the dismembered leg.

Then there was nothing left to do but wait. After two days, we removed the bag of sand from the second twig, but left it in place on the third until five full days had elapsed. The results were interesting. The first lichen, which had been consistently exposed to the light, stayed yellow—and, in fact, slightly yellower than before. It had been given more light than it had at the crime scene and had responded accordingly. The second lichen was yellowish with a greenish tinge, but the third twig, having been deprived of light for five days, was completely green. The ramifications for our case were clear. If this particular lichen turned green after five days of being covered up, and it was still mainly yellow at the deposition site, it meant that the leg could not have been placed over it for longer than five days.

When David and I reported our findings, the police could hardly believe what they were hearing; they had been convinced that the leg had been there for two weeks. The tighter time frame might have upended the accepted wisdom about the case but, as with lots of things in nature, the lichen was not lying. The investigation was not as neat and tidy as the police had supposed.

David and I continued to help with the investigation, now nicknamed "the jigsaw case," of the dismembered man. Eventually, the torso was found. It had been wrapped in a blue towel, put in a cheap suitcase, and dumped in a stream, miles from the limbs we had already examined. We waded in the freezing stream while the torso was recovered, but the best place for examination was the mortuary. This was only the third dead body David had ever encountered in his forensic experience, and his first time in a mortuary. I look back on that day with some regret because I was so insensitive. David is like me in so many other ways that I had assumed he would not be affected by the atmosphere and activities in a mortuary. But I can still see the pallor of his face, and his slight stutter, when I asked him if he would rather measure the fungal colonies on the torso or kill the maggots in hot water in preparation for the forensic entomologist. He opted for the maggots and I remember him being incredibly quiet in his little corner of the stainless steel bench with his kettle and bottles. Later, he confessed that he hated killing those wriggling little things, and especially killing the pretty beetle that had unfortunately joined the feeding throng inside the suitcase.

What began with a leg discovered on the edge of a roadside ended with the revelation of a story so vile that the most imaginative crime novelist might have struggled to make it up. After being invited to become a lodger at his friend's home, and growing envious of his new landlord's financial status, the killer and his girlfriend, a much younger prostitute, and the mother of two girls, conspired to kill him. Stabbing him in his back as he slept, the lodger, who had been a butcher in a previous life, and was reputed to dismember bodies for criminal gangs outside London, took to his old trade, carving up his former friend and scattering him far and wide. Perhaps they thought they would never be caught—but, as I have seen time and time again, the tiniest hint from nature can help point us in the right direction so that justice can be served.

CHAPTER 10

Last Breaths

Our bodies are dynamic and constantly changing. This is because they are being built up and broken down by our own biological processes all the time. As we breathe the air, and eat our food, we are taking the outside world into our own "inner sanctum" and, put simply, we take what we need and eject anything that is unwanted in the form of sweat, urine, and feces. What most people cannot realize is that there are tiny amounts of radioactivity in our food and water, and these get built into our soft tissues, bones, hair, and nails. Every part of the world has its own radioactive signature in the form of radioisotopes and, because of these signatures, we are able to trace your geographical movements since you were born. A tooth will identify your place of birth, a femur will tell us about your travels within about the last ten years because the bone is turned over every decade, and your hair and nails give information about where in the world you have visited more recently. It takes about one month for one-sixth of your fingernail, one-twelfth of your toenail, and about 1.3 centimeters of hair next to your scalp to grow. This means that information about your whereabouts, month by month, can be tracked.

The air provides the oxygen we need to release energy from our food but, as we breathe in and out, our bodies can retain traces of the geographical location we breathed in the air. As well as radioisotopes, the air is full of particles and debris and, if you are ever in any doubt about this, spare a thought for all those whose eyes start streaming, and noses start running, on a dry summer's day. Anyone who suffers from hay fever will attest to the fact that the air we breathe is full of pollen grains, plant and fungal spores, as well as other, unknown, allergens.

Just look at a sunbeam streaming through the window and you will notice that it is full of little specks, floating and swirling with any little disturbance. Most of us are unaware of this so-called "air spora"* we breathe in. After all, one of the prime purposes of the mucous membranes lining our noses is to trap little bits of foreign material and prevent them from penetrating too deeply into our sinuses and our lungs. But those of us with sensitivities or allergies certainly suffer the effects of these irritants. For considerable periods they can remain trapped on the membranes in our nasal cavities, especially on those covering the turbinate bones, the groove-like air passages that divide our nasal airways, and direct inhaled air to flow steadily into our lungs. We have no idea how

---

* Air spora: all the tiny particles floating in the air. These are usually pollen grains, spores, and fragments of organic material and dust.

long these particles can remain intact and experiments to find out would be virtually impossible. The opportunities to observe the actual pollen load of the turbinates are necessarily rare; there are, after all, only so many corpses to be stripped down and examined. Yet, thanks in particular to some of my cases, the screening of cadavers for the palynomorphs, inside as well as outside them, have gradually become acceptable to pathologists at postmortem examination. Crimes have been exposed, and justice has been served, by virtue of the things our bodies have accidentally acquired.

Go back in time twenty-five years, to the city of Magdeburg, sitting high upon the River Elbe in Saxony-Anhalt, Germany. Magdeburg has experienced the wild ups and downs of history more than most other settlements of its size and, in 1994, another piece of its grisly past was exposed. The foundations for a new apartment block were being excavated in the middle of the city when a mass grave, containing thirty-two unidentified skeletons, was unearthed. Based on witness reports, and the poor state of the dentition of the victims, they were thought to be Soviet soldiers; but the Magdeburg community was divided in its opinion on who was responsible for the deaths. There was every chance that, in the tumultuous spring of 1945, the Gestapo had carried out the mass killing, but the rival theory was that the agents of the Soviet intelligence agency, SMERSH, which had had its postwar headquarters in Magdeburg, probably murdered them while putting down a revolt in 1953. If the Gestapo had done it, it must have happened in the spring, whereas the revolt put down by SMERSH happened in the summer. Reinhard Szibor from the Otto von Guericke University in Magdeburg thought that if the season could be established, the riddle would be solved.

The turbinates are high up in the nasal passages and are formed from a veritable coral reef of very thin bone, covered by thin mucous membranes. The stickiness of the membranes means that any particle becomes trapped on them until the mucus is removed by nose-blowing. Szibor decided to check whether he could differentiate between spring and summer pollen trapped in the turbinates of the buried victims, and a yearlong experiment seemed to prove the validity of his theory. He asked one of his graduate students to blow his nose at regular intervals across a year, and to then identify the pollen grains in the series of handkerchiefs. The results led them to believe that they could, indeed, differentiate spring from summer pollen from the handkerchiefs. He considered that alder, hazel, willow, and juniper would represent spring, while rye, plantain, and lime would be more abundant in the summer. Based on his observations, and this one set of nasal observations, he came to the conclusion that the victims had met their end in the summer and consequently that, rather than the Gestapo, SMERSH had been responsible for these particular killings.

Pollen had been used to date archeological discoveries before—but Szibor's was the first to attempt its retrieval from the turbinate bones of skulls. He was convinced that he had demonstrated a difference between spring and summer by this method, and I was very impressed by this. But the BBC showed me a video of him demonstrating his technique and, although I had been invited to the studio to extol the virtue of the method on the popular program *Tomorrow's World*, the viewing in the green room before the program made me decidedly uneasy. After the viewing, I was more than skeptical of the perfect results that Szibor had claimed. Szibor was not a botanist and seemed to have ignored both contamination and the phenomenon of pollen residuality. If one season's pollen had been preserved in the Magdeburg soil for well over forty years, it was likely that pollen from all the seasons was there too. I have worked long and hard in soil palynology and, in some soils, preservation can be very good indeed—but the soil matrix is likely to contain the input for a whole year, and even pollen from previous years. Certainly soil animals, especially earthworms, and many tiny arthropods, mix huge amounts throughout the profile so that various seasons' pollen would be likely to be mixed up. I thus had to question how Szibor found one season's pollen on the turbinate bones that had been buried in soil for forty years, and were likely to have been grossly contaminated by it. He claimed excellent differentiation, but it is possible that his findings were fortuitous.

Nevertheless, in spite of my reservations, I thought this was a brilliant idea. In my view, it is utterly essential to eliminate contamination of the turbinate from any part of the skull, and the soil comparator samples would need to be thoroughly homogenized so that a true picture of its pollen load could be gained. From then on, whenever I had the opportunity, I flushed out cadavers' turbinates to see if it would give useful evidence. It certainly proved so in a few investigations and, in 2000, six years after my first tentative steps into the forensic arena, the technique helped to solve the riddle of a young man strangled in a Hampshire woodland.

On a bitter December day, two mornings after Christmas in the year 2000, a man was out with his dog, working off the effects of too many mince pies and enjoying the brisk winter air in an area of woodland about twelve miles northwest of Portsmouth. Nowadays it is mostly commercial woodlands but it still retains some vestiges of the ancient Royal Forest of Bere and, crisscrossed as it is by bridle ways and paths, it is some of the most accessible woodland in Hampshire. The man's dog suddenly scooted off and disappeared into the trees, and he did not respond to his master's whistling, which was very unusual for him. A little farther on at the edge of the path, the dense undergrowth gave way to a sloping area of turf, obviously kept short by rabbits and deer, and perennially wet, as evidenced by clumps of rushes and sedges poking through here and there. It was an obvious place to get into the trees without having to do combat with bramble and dead bracken.

The man stopped and listened. He could hear frustrated whimpering to his left—and that was where he found his dog in a little glade, scraping at a round object poking out of the soil beneath a huge beech tree. The man pulled his dog away and poked at the object with his walking stick but lurched back in fear when he saw that the round object had an ear. The head was facedown and the rest of the body was buried.

A large log had been placed over the grave. It is strange, but murderers often mark a grave in some way, possibly to make it easier to find. Many have been known to revisit the graves of victims, possibly to check whether the body is sufficiently hidden. But who really knows what goes on in their minds?

Soon, police officers arrived to set up the inner and outer cordons and some unfortunate constables were allocated to guard duty. The victim had been missing from his home in Portsmouth for more than six weeks by the time the police called a forensic archeologist to retrieve the body. His family had got through this Christmas, not knowing where their twenty-four-year-old son had gone and, after he was reported missing, no one was able to give them any answers. All that anyone knew was that he was last seen in his white Ford Escort van at Hilsea Lido in Portsmouth on 11 November. That same van, on the very same night, had been found on a local industrial estate, set ablaze and destroyed.

The dead man was definitely in a state of decay, but the low winter temperatures must have slowed the process so much that no standard decomposition model would have been able to establish how long he had lain there. The pathologist quickly established how he had been killed and, although there was a deep knife wound in his side, what had finished him off was tight garrotting. A cord had been put around his throat and using a stick at the back of the neck, the cord had been wound, around and around, until he was throttled. It was a brutal killing. It later came to light that he was murdered because he had taunted one of his "mates," obviously a man with a vicious temper and dominant enough to implicate an accomplice. The police also established that the victim had not exactly been the most law-abiding of citizens. A joiner by trade, detectives believed that he had carried out a number of burglaries on an exclusive housing development close to where he lived. By thorough, routine police work, two suspects were identified; one of them had been seen in the victim's van on the night he vanished. As a matter of routine protocol, footwear from both suspects, and a vehicle belonging to one of them, were seized as exhibits to support a prosecution case against them.

And that is where I came into the case.

Woodland has been a favored place to dispose of bodies since time immemorial, and there are woodlands in Britain that might even be regarded as mass graves. I was once involved in a case for the Metropolitan Police where they were searching a woodland for twenty-four victims of a gangland boss, the burials having occurred over a number of years. In the Hampshire case, the sinister place holding the victim's grave would under normal circumstances have been a most delightful glade. There were several large beeches, but also holly, hazel, and wild cherry. There were banks of bramble at the edges and the gnarled, twisting stems of old honeysuckle plants were winding and grasping at the stout supports offered by shrubs and trees. Much of the ground was covered by glossy ivy, which had managed to creep up some tree trunks in its quest for light. This was winter and the ground was rather bare except for thick layers of beech leaves, beech nuts, and acorns, and, although there were many trees bordering the glade, a bridle path could be seen through the gaps in the winter branches.

The question the Senior Investigating Officer (SIO) wanted answered was whether the victim had been killed in the wood or if there were, in fact, two crime scenes. Had he been transported to his grave or had he been living when he came to this glade and killed right here where we stood? We also had to ascertain whether the shoes belonging to the suspects had contacted the scene. As fortune would have it the police had quickly identified two suspects, and they gave me their footwear, along with the floor mats and foot pedals of the main suspect's car. Indeed, one can only claim "shoes" and not point to a person because of the habit that criminals have of sharing footwear; because of this I needed to be able to eliminate as many other places as possible as being sources of the profile that I had retrieved from the suspects' belongings. One good thing about this case was that both parties were very much "city" boys and very unlikely candidates for taking walks in the woods for pleasure. That, of course, would not prevent any defense attorney claiming that they did just that, so I had to be prepared.

The whole area, including the public car park about a third of a mile away, had been closed to walkers and visitors. I walked from the car to the grave with John Ford, the tough police sergeant who was responsible for the day-to-day running of the case, and I was pleasantly surprised to discover that he was genuinely interested in what I was doing and wanted to learn about vegetation and plants. He had a broad Hampshire accent which made him seem homely, but I soon learned how tough he could be. He was determined to gather every scrap of evidence that might offer itself. I liked him. He was open and blunt, and I knew I could work with him.

As we walked along the path created by estate workers for the benefit of public access and enjoyment, I marveled at the richness of the landscape around us. This was a huge area of woodland—a patchwork of woodlands within woodlands, each with its own character and species profile. There were obvious plantings of conifers and dense stands of birch, but there were also magnificent old beeches and oaks, with an understory of hazel, holly, and a tangle of the bare stems of many species defying immediate identification in this winter light. There were many stands of sweet chestnut trees that in antiquity provided essential food in southern Europe and were brought to Britain by the Romans. There were also individual trees of one of our four native conifers—the dark, brooding yew. One patchwork piece of woodland merged into the next, but the herbs, which would open up into another kaleidoscope of colors in the spring, could only be predicted at this point because they were in their winter slumber with little showing aboveground.

As I walked along, I was envisaging the kinds of botanical profile each area might yield. But one thing I have learned, and novices still cannot grasp, is that one can never be sure what will be revealed. I could make broad predictions—I knew that there would be high levels of pine and birch pollen near the car park, that sweet chestnut would hardly register because it produces relatively little pollen, and that there would be more oak than beech, even though beech trees were numerous. But the exact quantities and patterning? Analysis would be needed to reveal this and no modeling exercise would give results that could ever stand scrutiny in court. Every case is unique and must be treated as such.

After all my training and experience in archeology and ancient landscape reconstruction, one thing that originally and still continues to surprise me is that every sample I collected from the surface soil would be different from the next and, the farther I sampled away from the first, the greater the difference would become. In fact, everywhere, pollen fallout is patchy and only moderately predictable. Its patterning can be seen in terms of spectra, one spectrum merging into another as the vegetation changes. As I have mentioned a number of times, when comparing objects and places it is essential to collect sufficient comparator samples to enable the picture of place to be constructed. It also helps enormously if some rare speck of trace evidence makes that place particularly distinctive. Nothing must be left to chance if one is to give a robust performance in court—for performance it certainly is. Half-baked predictions in the absence of hard evidence would soon be torn to shreds if the lawyer on the opposite side were worth his salt. I was also aware that I had to be strict with myself so that I would not be infected by the innocent enthusiasm of police officers for the desired outcome—conviction. I am always conscious that one must constantly fight cognitive bias.

No palynological analysis can ever offer absolute proof of contact. Everything must be viewed in terms of likelihood and, every time I make a recommendation or report to the police, I am careful to add the appropriate caveats. One must look for alternative scenarios that might account for any profile obtained from a suspect; and this essentially was to lead my police companion and me a merry dance across Hampshire and West Sussex.

No one could have reached the grave site by car, and the victim would have had to be walked, or carried, to his death. This meant that I needed to identify the easiest way in and out of the site from the road, and to target my samples in any place likely to have been contacted by the offenders. The main and obvious one, of course, was the grave itself; that is one place they could not avoid. I was given the footwear, floor mats, and foot pedals of the main suspect's car for analysis, and my word, there were close comparisons with the samples from the crime scene itself; and importantly there were some remarkable specific markers. All the trees and shrubs I had seen were very well represented and, in the spring, the place must have been exquisite. The pollen and spores showed that the site had lots of bluebells, wood anemones, dog's mercury, ferns, and other herbs that are characteristic of such places, but most remarkable was something I had never, ever found before—Solomon's seal. My microscope also picked out some peculiar fungal spores that looked like the legs on the Isle of Man flag, without the "knees" being bent. These turned out to be the spores of *Triposporium elegans*, a microfungus that infects beech nuts particularly. Not only all that, but I also registered a hay meadow assemblage right in the middle of this quite dense woodland. That might have presented a bit of a conundrum if it had not been for my experience in that Yorkshire cellar, and the fact that I had noted a bridle path near the grave site. It was obvious to me that the hay meadow profile had come from the horse dung. This gave the crime scene a high degree of specificity—a signal of hay meadow within deep woodland.

The pathways in and out of the grave site were important, but so was anywhere else frequented by the suspects. This led to lengthy visits with John, my police sergeant companion, to the suspects' homes and all their favorite haunts. The experiences proved to be arduous but not without humor, especially when visiting various addresses in Portsmouth itself. We needed to check out the palynological status of a number of modest terraced houses in some less-favored parts of the city, and the contrast of one with another was astonishing. I clearly remember the first one. Walking through the dark narrow hall, avoiding the huge jumble of coats on hooks just inside the front door, really sticks in my mind. It opened out into a living room that could only be described as chilly and sparse. There was no hint of homeliness, although the baby crawling on the threadbare carpet looked bonny and clean. The room led through a scullery, then on to an entirely neglected backyard, and an excuse for a garden. There were old buckets, neglected toys, and old boots in the backyard, and some of the weeds, including docks and nettles, growing between the paving stones were about 30 centimeters tall. A dejected clothing line with abandoned pegs hung slackly along from the house to a scruffy shed, and the whole dwelling left me feeling sad.

Not so the next one. It was a two-up, two-down terraced house like the other one, but that was where the similarity ended. This one had new double-glazed windows and had been extended out toward the garden. A woman opened the door in a cloud of a strong, woody perfume and immediately one could tell she had money. She was expertly made up, with hair too blond to be blond, and smoking a cigarette. She was wearing black leather trousers, a black sweater, and several bright gold chains, neck to waist. Pink toenails peeped from her glittery mules and her long fingernails had been manicured into fashionable squares with white-painted tips. Several rings adorned her hands, including a large diamond cluster. The décor was a revelation—lush red carpet, black leather sofa and chair with fur cushions, and an elaborate crystal chandelier slung from the ceiling. The TV was as huge as a cinema screen and the little bar that had been squeezed into a corner was groaning with bottles of booze and glasses. The modernized fitted kitchen led outside into a sterile space of immaculately clean concrete slabs, with one or two tubs of dead flowers from last summer. There was nothing at either of these houses that resembled the vegetation where the body had been found. I had not expected there to be and could tell at a glance that there was no need of comparator samples and analysis from these places.

I still had to counteract the inevitable claim by defense counsels that their clients regularly walked through woodlands all over Hampshire and Sussex. When you think about it pragmatically, and from your own experience, how many woodlands are likely to give a profile that would include those specific trees, shrubs, climbers, herbs, peculiar fungal spores and, on top of that, Solomon's seal and a hay meadow, all in those proportions? From all the years of examining ground surface samples from a very wide range of places in the UK, I had never seen anything like it, and my gut feeling is that one does not exist. But one must play the court game of plugging holes of doubt.

I was not familiar with the woodlands in the defined area but I knew a man who was. My teacher and continuing mentor from King's College, London, Dr. Francis Rose, MBE, whom most British botanists revered because of his encyclopedic field knowledge, was the obvious oracle to consult. I took John with me. By this time, I had already analyzed the footwear and vehicle and had produced profiles that were very close to those of the comparator samples. I said, "Francis, do you know of any woodlands in this prescribed area that could yield these profiles?" I gave him my species lists and tables of proportional data. He sat back in his chair in his sitting room, which was stuffed with books, the table spread with plant specimens, pencils, his lens, and notebooks. He chewed the end of his pipe, scratched his beard, peered over his glasses in his usual, smiley, avuncular way, and stretched out to a bookshelf bearing lots of well-worn maps. After about an hour, and several cups of tea while squinting at Ordnance Survey maps covered in circles and scribbles, he picked out fourteen areas of woodland that he knew could be relevant. My heart sank at the enormity of the task ahead, whereas John, who had been transfixed by Francis, simply said, "Well, that's it, then. We'll start tomorrow, Pat." We did too.

We visited all fourteen woodlands but, just by looking at the vegetation of many of them, I was able to eliminate most, so the task was not as bad as it first seemed. They all had oak, beech, pine, and many of the other plants in my profiles, but only three came near to the plant community at the grave site. Even then, after analysis, none of the pollen and spore patterns gave a true picture of what I found at the grave, and what I had gleaned from the footwear and car. The unusual Solomon's seal and the spectacular fungal spores were not found in the other sites either. I had thus satisfied myself that there was a strong similarity between the suspects and the grave site, whereas there were only weak similarities with the other sites. But an important question had yet to be answered: Had the victim been alive or dead when he was taken to that hole in the ground in that pretty glade of doom? I needed to work on his body to see what information it would reveal.

There was no point in washing the hair because it had been exposed to the grave soil, but the turbinates might give us something. I arrived at the mortuary to find two glum-faced officers.

"Sorry, Pat," one of them said, "the mortuary staff put the cadaver in the freezer—he's as solid as a frozen chicken."

They were anticipating my frowns of disbelief—why hadn't they telephoned me before I left home?—but instead I simply asked, "Have you got a hair dryer?" and two were on the table within fifteen minutes. We all took turns in blowing hot air onto the victim's skull and face. We were all bored because we knew it would take some time to defrost the skull, even though the brain had already been removed. I recall the macabre humor of the detective constable staring at me, with a poker face:

"What did you do at work today, Daddy?"

I nearly fell off my stool at this but, of course, jokes are not encouraged. The deceased must be treated with every respect and their dignity must be protected. We pulled ourselves together and, eventually, the head was in a good enough state for me to begin flushing out the turbinate bones. Obtaining palynomorphs from the body of a murder victim is no simple procedure.

Picture, for a moment, the victim's body laid out on a laboratory slab. In my earliest attempts to obtain palynological remains from a victim's nasal cavity, I would enter through the nostrils, with a flexible tube attached to a large syringe of hot, dilute detergent water, reaching up and into the turbinates to flush the, hopefully, palyniferous membranes. The nose is only the visible part of a whole respiratory tract, and the nostrils lead directly into a cavity separated into two by a vertical partition, the septum. The nose's principle function is to warm and moisten the air before it gets drawn down into our lungs, and to filter out any foreign particles along the way. Nostril hairs, the bane of so many aging men and disgust of so many women, help to prevent foreign material getting too far, but it is the large blood supply to the turbinate membranes that warms the inhaled air from the nostrils, while the layer of tiny beating cilia on the membrane surface trap and move any alien particles back down toward the nostrils, preventing them from entering the airways. Mucus is secreted throughout the respiratory tract from nostril to lung and this also helps to trap foreign particles. Yet, accessing the turbinates by the nostrils is awkward; and too often, unless the face and nostrils are impeccably clean, flushing with a catheter can result in the collecting water picking up contaminants. The nostrils can pick up any dirt and particulates, especially for a corpse where decay or putrefaction has set in, and the risk of contamination is high; and even the pathologist, in his routine washing of the body, can swill contaminants up into the nasal cavities.

By now, I had modified and refined Szibor's technique. In previous cases, wherever I could, I had been removing the nose, and sometimes the whole face so that I could get rid of extraneous material before flushing with my syringe. But I was still not happy. The method was too crude for my liking and, on advice from Sue Black, anatomist and anthropologist, I approached the turbinates in a different way.

The body had been facedown in its woodland grave for as much as six weeks, and the whole head was naturally rimed with the soil and decaying leaf mulch that had filled the hole. Instead of going in through the nostrils and risking contaminating whatever I might find, I would take the advice I had been given and, flush the turbinates from the cribriform plate. The first time I ever saw this special plate of bone, perforated with little holes to allow the olfactory nerves to pass into the brain, I admired evolution. What a perfect little structure, sitting above the nasal cavity and separating it from the frontal lobes of the brain. To see it, however, meant that the top of the skull had to be removed as well as the skin of the head and face; these had been put back by the pathologist after removing the brain for examination. It is usually an easy job because it has already been removed once by the pathologist at the postmortem, and it merely involves gently lifting it away from the bone. At the end of my work, I could easily replace it and no one would be the wiser.

The most difficult part is always orienting the body such that the holes of the nasal passages are placed directly over my stainless steel kidney bowl, one of my prized possessions that I have had for over twenty years. Getting the corpse into position usually takes a great deal of effort and I invariably need the help of the mortuary technicians. I block off the throat with a wad of nonabsorbent cotton wool, then, punching a hole on the left side of the cribriform plate with a scalpel, I insert a syringe with about 20 milliliters of hot antimicrobial detergent solution (nothing more than my faithful servant, medicated shampoo) into the hole. As I gently flush, the solution gushes through the coral reef of delicate turbinate bones, taking any adhering particles with it, out into the bowl. I then repeat the process on the other side of the plate and pool the two sets of flushings into a single suspension. After mixing, the sample is divided into two, as insurance against one being lost, and both tubes are centrifuged to get a pellet of the particulates as plugs in the bottom of each one.

Sometimes the particulates recovered are in such small quantities that they are barely visible to the naked eye in the bottom of the test tube—but, even then, the samples must be processed and studied. I have known cases where single particles have provided a vital point of difference, changing the complexion of how we understand the story. The small stuff matters—these samples represent the last breaths taken by the victim and might tell us where that was.

The sample I had recovered from the victim's turbinates was richer than I had imagined. My expectations had not been high. The number of palynomorphs retrieved from nasal passages are usually rather low; in many examinations I have conducted, cadavers have yielded fewer than ten grains for every 40 milliliters of solution washed through the turbinates. And yet, as I worked through the flushings I had recovered, the number kept rising—until, at the final count, we had identified 739 separate palynomorphs, belonging to 35 palynomorph taxa. What is more, the profile from the turbinates was very similar to the surface soil around the grave—but certainly not like that of the soil deeper down.

Soil is remarkable. It is composed of part mineral and part organic material, and it is teeming alive with bacteria, fungi, and animals. Most of these are active in the top centimeter and, as the soil becomes deeper, both the number of organisms and their activity get less and less. At a depth of about 8 centimeters, the number of viable organisms drops away dramatically, and even more at about 20 centimeters. It is the same with palynomorphs. They are relatively abundant at the surface but, deeper down, they have been in the soil much longer with much more opportunity to decay. Thus, with depth, the number of pollen taxa usually diminishes and the grains themselves become thinned, corroded, and crushed. What I retrieved from the victim was all very well preserved—which meant it came from soil closer to the surface; this represented very recent pollen, probably that of the previous year.

For soil particles to have been breathed up as far as his turbinate bones, he must have been breathing very heavily indeed. Just imagine having your face pushed into the soil surface and trying to breathe. You would eventually be gasping through your mouth and nose, inhaling with animal urgency. Now I was certain:

this young man was struggling for breath as he was being gar-rotted. With his nose in the soil, he would not have been able to avoid breathing it high into his nasal passage, well beyond the nos-trils. The pollen and spores could not have got there by taking a stroll through the woods, and if he had been garrotted back in Portsmouth, he probably would not have inhaled anything like the number. In any case, the pollen and spores in the nasal passages closely resembled the soil on the surface of the ground. I was near certain of it now: the woodland was the place of his murder. There was no other crime scene for the police to worry about.

Sometime later, with both killers convicted and seeking reduc-tions in their sentences, the confessions began. Each blamed the other for the murder. Each claimed that he had stood on in horror as the cord went around the man's neck and the last breaths were choked out of him. The killer might have been sufficiently forensi-cally aware to drive his van to a lonely spot and set it alight, de-stroying any evidence associated with it, but they were not aware enough to know that, as they dragged him into the woodland, it was leaving its own impression on them; and they did not know that, as the poor man lay there, with the garrotte around his neck and his face pressed into the graveside soil, he was breathing in the very particles of pollen and spores that would one day see them imprisoned for life.

This case came up when I was teaching the Master's in Archeo-logical Forensic Science that I had proposed, set up, and coordinated at University College London, actually in the Institute of Arche-ology. Forensic archeology had been popular for some time, and Bournemouth University had set up a master's course in the subject, with many young archeologists desperate to get involved with criminal investigations.

I had taken up early retirement from the Institute because of increasingly poor health. Working out on sites in all weathers had proved just too much. I remember standing up to my thighs in a ditch full of freezing water one Christmas Eve. It was getting dark and I was shivering and yet hot. My head ached, my back ached, and it began to hurt when I breathed. By the time I got home, I was in the depths of pneumonia and it took me a very long time to recover. My doctor said that enough was enough, and that I could not go on punishing myself any longer.

"You'll just have to retire," he said.

"What?" I gasped. "I can't do that, I have just too much to do!"

I was distraught as I had, by now, a considerable number of very successful forensic cases behind me, and I was still learning and developing the skills that were needed to do my forensic discipline, and the courts, justice.

The director of the Institute was a difficult man but he always seemed sympathetic toward me.

"I'll do you a deal," he said. "You put on a master's course and I'll give you the lab and all the facilities to develop your forensic work. Just don't go expecting a salary . . ."

I thought that was a fantastic offer because I could be paid for any police work, it would not involve the heavy physical endurance needed for archeological fieldwork, and I would have the assistance and backup I needed. That is how the course came into being, and I look back on it mostly with pleasure—teaching the subjects that were close to my heart, and still doing casework. My fate seemed to be a teacher yet again.

I was determined to make my course very wide-ranging and I

involved specialists from many other fields to make it come alive. I set up a rota and the students took turns in accompanying me to crime scenes and the mortuary. That certainly sorted out the sheep from the goats; I was amused that it was the women who excelled, the males more often being rather feeble when it came to real-life experiences; well, real-death ones actually. I could write another book on my experiences running that degree course, about all the marvelous young people who followed me and the delights and disappointments that were the companions of teaching so intensely. My students had been exposed to the usual lectures and practical classes, but also to crime scenes, police stations, mortuaries, *postmortems,* and courts. It was certainly a varied and comprehensive course.

CHAPTER 11

"An Empty Vessel"

*B*ehind *every strong and independent woman lies a broken little girl who had to learn how to get back up and never depend on anyone too much."* I do not know who wrote that—I just saw it on the internet—but it is so true in my case.

Life in our village went on, as it had done for so many years before. My grandmother still stayed with us on her itinerant rounds. My lungs were still weak, giving in as frequently as ever to bouts of pneumonia, pleurisy, and bronchitis. My best friend and I had drifted away from each other in recent years, sent off to different grammar schools after the placement exams, her to one in Monmouthshire close to our village, and me to one in foreign Glamorganshire. The school bus wound its convoluted way down the valley and around the villages to take us on the long ride to the drop-off on our side of the county divide. Never mind the weather, we then had to walk down a steep, winding road, cross over the Rhymney River bridge, and climb the longest, steepest hill to the strictest, most forbidding place one could ever imagine—Lewis School for Girls. We never questioned the hardship of that slogging trek once we got off the bus; we just endured it as children never would today.

I soon made other friends including ones from Glamorganshire, most of us children of the coal-mining culture, whether our fathers were "real men" at the coalface or worked in the mines in some other capacity. The only concept of any class hierarchy we had was that gained by exam results. The community was passionate about education and fathers were blessed with the Miners' Welfare Institutes to spend hours of quiet in the reading rooms. My father was one of the best-read people I knew and he could always hold a cogent argument on just about any topic. He taught me to argue too, and my mother found it impossible to cope with either of us when we were discussing some issue or other out of the newspaper. Argument is a good game, rather like fencing—thrust and riposte—the main rule being objectivity and detachment from the subject. I have always enjoyed listening to measured, intellectual argument, just as baroque music and the Dutch masters are my favorite art forms. Precision and detail have always hypnotized me.

The teachers at my grammar school had an easy time—old bats. I hated that school. It was as loathsome to me as my junior school had been delightful. All the girls (there were no boys—hence the delights of chapel, where there were plenty) were bright, and the black-gowned teachers never had to experience insubordination or difficulty in conveying information and concepts. There were some compensations for the girls, though. The camaraderie that existed was strong and real, and a group of about fourteen of us still meet up in Cardiff Bay each year to exchange news, views, and opinions. I now know that I was not the only one who was made to feel inadequate and whose spirit was so dampened by the oppressive discipline. There was little nurturing of talent and, looking back, the attitudes were positively Dickensian. But there were some highlights; every year in early March the school buzzed with enthusiasm for our annual Eisteddfod, held to honor culture in all its manifestations. Every girl belonged to a house, each one named after a local mountain, except for Lewis house, named after our founder. I was in Bedwellty house, and our color was yellow. To this day I can still remember which girl belonged to which house because during Eisteddfod week, they were either collaborative friend or competitive foe.

The ethos of that school was excellence in all things, and we all had to be obedient and work hard, or take humiliating punishments. One nasty penalty was having to learn poetry by rote overnight and then recite it in front of a whole class of peers. I still cannot learn anything off by heart, and I have no love of poetry—it is still a punishment to me even though I accept the depth of beauty and meaning in some of it. Needless to say, I never chose to recite poems at the Eisteddfod. Everyone had to perform, whether it was recitation, singing, dancing, painting pictures, or writing. To be fair, every conceivable activity was included so that each girl had the chance to excel at something, even if it was growing daffodil bulbs in a bowl. But life was definitely competitive at every stage, very full, and hurtling forward to the big exams, just weeks away.

It was a Wednesday night in April 1958, just before my O-level exams, and I was enjoying myself at a social event in the Baptist chapel. I loved chapel and attended twice on Sunday and, sometimes, on a Wednesday too. We had great fun and the best-looking boys went there, so there were other incentives besides worship.

That Wednesday night, my world came up against a brick wall—it was punctured and the air hissed out of it. A friend of my mother's had come to the chapel hall and wanted to see me outside. I was puzzled.

"Look, Pat, don't go home tonight. You're to go to Auntie May's. Your mother's left your father, and that's all there is to it."

I stood stock-still. "What?"

"I'll take you up to your auntie's. She is expecting you."

Wrenching a child unsuspectingly from her home in such a brutal way is less likely to happen today; society has been educated to care about the emotional well-being of their children. But, I had gone to chapel in my green dress while, unbeknown to me, my navy-blue-and-white school uniform, and my school books, were being delivered to my great aunt and uncle's home a few miles up the valley. Ever since I could remember, I had never liked living with my mother or my father. They were volatile, they were too passionate, and they constantly argued and fought over the most trivial things.

Theirs was a war of personalities rather than intellect, and I was invariably caught in the crossfire. But my mother leaving my father? That was unthinkable and in that day and age, families just did not break up. They stuck it out, and the bonds of marriage were bonds for life. All I could think about was the shame that would follow the announcement. My cheeks were on fire and I felt sick, and insecure, and frightened. Why couldn't my mother come to fetch me? Why was it so sudden? Why did I have to be farmed out to elderly relatives who never struck me as being particularly sympathetic toward us? They were childless; their house was spick and span, highly polished and cold.

I did not speak again, not even when I was delivered to my new, and hopefully, temporary home. I did not ask about my mother; I did not ask about my father. The truth was that I was ashamed, humiliated, and disgusted at the behavior of so-called adults. I always felt more grown-up than either of my parents, and I was used to viewing them through critical eyes. As ever, they were putting themselves first—their feelings were the most important, and had to be considered at all costs, even at the expense of their children. And yet part of me was glad that my parents had parted. Love is not always enough, and the combination of the two had been toxic for too many years.

To be relieved of being caught in the middle of their war was, in many ways, a blessing—but it came with its curses as well. At sixteen years old, I felt the stigma of their separation like a brand. And I will never forget or forgive my mother insisting that I accompany her to the court and being told that, from this moment on, I was to live with her. Looking back on this time, my parents did everything that is now considered to be bad for a child's psyche and well-being. These days, it would be classified as child abuse. Then, such a concept would be unimaginable for a girl from such a "good" home. The worst thing was that my grandmother was absent and I had no sanctuary. It was the turn of her youngest son and his family to have her, and she was in Yorkshire.

I confided, in a whisper, to a special friend about my parents' separation and, as her parents had been engaged in a cold war ever since she could remember, she was sympathetic. I have always considered parents and teachers to be the most influential, and the most dangerous, people on the planet, and I felt that I was a victim of both. Parents and teachers can destroy, they can enhance, they have the power to create a victim or a protégé, and the children within their sphere of influence have no power at all. Well, not in those days anyway.

The flat my mother moved me into was not home. It never would be. Home was the house where I had a room with my books and privacy. Home was the house where my grandmother would come to stay, and read alongside me, and nurse me when I was unwell. This new life did not look promising and, if I ever thought that my mother would soften when she was no longer fighting with my father, I was mistaken. Perhaps volatility had become a habit to her; she was as domineering and explosive as ever. Everything had to be done her way—and what this meant, most of all, was that I was never to see my father. His life continued, forever skirting the edges of our own, but my mother forbade me to have any contact; she forbade me to ask questions about him, or even to speak of him in her presence. This was not because he was a bad person but because he had committed the unforgivable—he had been unfaithful.

These days, this would certainly not be grounds to keep a child from its father, and I knew he was desperate to see me, that he kept on trying but, from that moment on, my mother behaved as though he had never existed. I still know nothing about the catalyst that caused their final separation but then I have never wanted to be enlightened. Now I look back on all this with disbelief that it was allowed to happen. Poor Dad with his wonderful hair and good looks. He had charisma too, but I could only cope with one of these volatile people, and as it had to be my mother, he became a remote, rather ghostly figure in my life. Parents beware: if you try to denigrate the other one to a child, you will forever be a traitor and less loved for it.

When my mother had succumbed to old age and was living alone in the huge Georgian house on the top of Bedwellty Mountain, surrounded by stunning views and sheep, the doctors told me that she simply needed care and not hospital.

"I am not going into any care home and be stuck in a corner like an old granny."

I knew that if any carer moved in with her, they would not last five minutes, and I would forever be trying to find responsible people.

There was only one thing for it. She came to live with me in my big house in Surrey, where I lived alone with Mickey, my beloved cat. So began three hellish months. She had always had someone around to do her bidding and, in the beginning, I was her handmaiden. Not for long—I was working very hard earning my living, and I now had her to care for, do her laundry, and make sure she had her meals on time. I became very tired:

"Where are you going?" "How long are you going to be?" "Who was that on the phone?"

It was déjà vu. I was being treated like a teenager all over again. But, gradually, she came to realize that I was quite a sympathetic person, and that I had a breadth of knowledge that she lacked. She witnessed some of my discussions with police and accompanied me when I lectured to the police training college at Hendon. She met some of my friends, was invited to parties along with me, and she began to realize that there was life outside the Sirhowy Valley, that tiny pond where she had been such a big fish.

After about three months of purgatorial existence, she slowly began to realize that I was kind and that she was comfortable; I realized that she was clever and very, very funny. So this is why she had so many friends and acquaintances! She often made me laugh out loud with her impersonations of people we knew and the things she said. She was still maddening and could be hurtful, but there was some depth there too. She could play the piano by ear, did exquisite embroidery, was brilliant at craft, she was pretty, could be very engaging, and people were drawn to her. I was, perhaps for the first time, getting to know and understand my mother a little. We had three months getting on quite well and the quarreling no longer characterized the relationship. Then, after she had been with me for six months, I had to go to New Zealand to a conference so I took my mother to Wales to a nursing home for the duration of my absence. We had arranged that she would come back to me when I got back from New Zealand but she broke her pelvis the day I got back and she rapidly went downhill from there on.

After nearly six months, on the day after Boxing Day, she died in my arms and her dying words were, "I never realized I had such a wonderful daughter." I wept, not because she was gone, but for all the lost years when we could have been friends. I do not think she could forgive me for being my grandmother's favorite, and being bestowed with all the love and attention she felt was hers. Her poor mother had always been too busy making sure her family would survive, and making time for open expressions of affection was probably a step too far. But I could never forgive my mother for her carelessness in causing my injury, the apparent disregard for my feelings throughout my life, and her constant invasion of my mental privacy. I could never relax in her presence. How sad it all was.

After my parents' divorce, life with my mother was utterly painful and, when my boyfriend said he was getting a job in England, it dawned on me that I could do the same. So, halfway through my A-levels, I applied for a laboratory job in the Civil Service in Surrey, got it, and found lodgings. Of course my mother tried to prevent it, but she just had to give in because I had made up my mind and was exhilarated at the prospect of my newfound freedom from her. I just ran away.

Looking back on it, the people with whom I lived in Surrey were sweet and kind but, oh goodness, they were so foreign. I was fed wholesome casseroles and had a nice, warm room but I was homesick. Perhaps that is hard to believe after what I have said about my life at home, but I was homesick for the valley, my friends, and all things familiar. I could not believe how flat it was in Surrey—I had not seen one proper hill. The water tasted salty and horrible; it was not the sweet, soft water coming from the hills. The buses failed to stop for you unless you put out your hand. In Wales, if you were standing at a bus stop, it was deemed obvious that you were waiting to be picked up. I was dumbfounded that bus after bus just drove past until I saw someone hail at the opposite bus stop and the thing came to a halt. This was another lesson in coping with this strange, flat land, where everything looked the same and there were rows of poplar trees everywhere.

I was surprised to see bottles of pickles and jam with the price stuck on the side, and biscuits in packets. I had never seen anything prepackaged and I could not help thinking it was really vulgar. At home, cheese was cut with a wire, biscuits were always loose in big tins, sweets in big jars, and bacon sliced on a machine with a lethal, rotating blade. Health and safety? It did not exist. Most amazing to me was that in England, coal was bought from a coal yard and came in small sacks. I had only ever witnessed coal being dumped by the ton in the road outside the house, some chunks being the size of armchairs. Fathers broke them up and labored to carry the gleaming anthracite to the coal house, before sweeping, then washing the street with buckets of soapy water. Neighbor helped neighbor at such times, and there was always the gift of beer at the end of the job.

I remember watching my father break up the coal with a sledgehammer, and both of us looking for fossil plants in the depths of the boulders. We found a lot too—mostly ferns but also some gigantic forms of stems that I now know to be the ancestors of our little horsetails. Thinking back to those days, we had fun on bonfire night with firework parties on the plot of land my father owned at the back of the house. Particularly memorable was the year when one jealous little boy threw a sparkler into the huge box of fireworks being boastfully held aloft by another. The pyrotechnics were spectacular—rockets whooshing along the ground in every direction, firecrackers jumping up girls' skirts, Catherine wheels whirling aimlessly and frantically around and around on the ground, and the loud cracking and banging accompanied by genuinely fearful shouts and screams—everyone running in different directions. Looking back on it, it was hilarious and proved what they said in chapel: you must not have jealousy or boastfulness—what a good demonstration of the rewards of both these sins. Others kept chickens on their plots, next door kept geese, and I still remember my father eating a goose-egg omelet. It was huge. Farther down the street, the land might be used to keep a pig. I can still hear the screams when they were dragged off to slaughter, and the frantic clucking, flurrying, and screeching when one of the chickens next door was having its neck wrung for dinner. I hated all that but we did have fun on our plot. I think my father kept it just for us to play on it.

The few years after running away from Wales were unhappy, miserable, and lonely, but I was determined that there was no going back. I never did. People made fun of my singsong voice and vowel sounds, and felt free to denigrate Wales and the Welsh whenever they wished. This was not bias against the Welsh—demeaning jokes about the Irish, Scots, and northerners, and even those from the West Country were just as acceptable. I gradually came to learn that indigenous people of the Home Counties are a breed apart, although they now moderate their behavior because of political correctness. The last half century has changed the whole country. Now that I have spent most of my life in Surrey, I have become fundamentally anglicized and, when I go back to Wales, I feel a bit of a foreigner—I am acutely aware of the strong accent that I once must have had, and the differences in attitude and culture. I have lived most of my life in limbo—Welsh to the English and English to the Welsh. That is the fate of an immigrant.

I married a very tall and handsome Englishman whose childhood could not have been more different from mine. His parents, their house, and their lifestyle all reminded me of an Ovaltine advertisement: Mummy and Daddy around a cozy fire in a comfortable, prewar villa, all in their checked dressing gowns drinking Ovaltine before bed. I can still hear the jingle of the advert. To me, his childhood seemed to have been like something out of an Enid Blyton novel, and his parents' attitude to life, the world, and the universe was classically conventional. Mummy really did stay at home baking in the kitchen while Daddy went up to town in his bowler and with his furled umbrella, completing *The Times* crossword by the time his train pulled into Waterloo station.

This apparent idyll was marred by the war and evacuation to the country and, for five years, my father-in-law saw his wife and son rarely. He was a very senior civil servant with important war duties and had been sent to Manchester. Gradually, it emerged over many years of little snippets and revealing remarks that family life had not been all it seemed. My husband had little rapport with his father and spent most of his time with his mother, who adored him. I actually got on with his father rather well and was not afraid to prick his bubble of pomposity from time to time. I remember my mother inviting them both to her house on her hillside for a holiday, and my father-in-law expressing utter amazement at how clean everything was in the Valleys, how beautiful were the views, and generous and funny the people. From anyone else, his comments might have seemed patronizing but I knew he was genuinely surprised that we were not all covered in coal dust. I remember my husband saying to me, "You are the daughter he always wanted," and, with a curl of the lip, "you being academic." Again, was there jealousy for affection innocently received?

My husband's father had smoked heavily for most of his life and, just after retirement, he sank into the inevitable blocked artery/diabetes syndrome that took both his legs. He died aged seventy-two, which we now consider to be quite young; modern medicine seems to have given us hope eternal where sixty is now the new forty. I remember sitting next to him as he lay on his deathbed, immaculate as ever with neat pajamas and perfectly trimmed hair and mustache. I held his hand and absentmindedly put my finger on his pulse. I was interested by the irregularity of its rhythm and strength. The last rapid quivers coincided with his giving a single gasp and staring me full in the face. With this last breath all animation left him.

I was fascinated by the change. He was Dad—and then just a body; this happened within a blink. I had never witnessed death like this. It made me realize that the spirit had left, leaving an empty vessel.

Years before, I had become a reductionist—convinced that the soul, the spirit, the being is only a complex set of physico-chemical reactions. We are all victims of our brain chemistry and our personal experiences. Whether your nature is saintly or that of a psychopath is largely out of your control. You can only ever hope to moderate your behavior—your thoughts are your own and peculiar to you. Intimately witnessing the very end of a life was, for me, very comforting. I was convinced now that he would not suffer anymore. There was nothing left to respond to anything that might cause pain.

About thirty years later, in 2005, my mother died in my arms, just as my ex-father-in-law had done. The sudden switch from life into death was the same dramatic metamorphosis, but my almost detached interest in the demise of two old people was that of a curious scientist. I have only experienced the agony of real bereavement for two other people, and every cat I have had the good fortune to cherish.

It is hard to believe that, at the time we divorced, I had been married for forty-two years, and yet I knew my husband little better than after our very formal, gentle courtship of nearly five years. For decades, emotionally we had been a pair of satellites revolving around each other but rarely touching. He was a hobbies man and was an excellent photographer, then scuba diver, then pilot of fixed-wing aircraft and helicopters, as well as being a good horseman. He later became obsessed and highly competent with computers. He always had to have the very best equipment, and no one else was allowed to touch any of it. I thought we were very well off and, at one time, we owned a Porsche, then a Ferrari—and I loved driving both. We had a twin-engine, eight-seater Cessna light aircraft, two horses, goodness knows how many computers, and other various electronic wizardry. Ever cautious, I learned to fly because I was terrified of being alone at five thousand feet with a heart-attack victim slumped over the instruments. We flew all over Europe at whim, stayed in lovely hotels—Cannes was our favorite resort—and no wonder many of our friends shared my illusion that we were wealthy.

At the age of sixty, when I was looking forward to a comfortable and fruitful retirement, the ugly truth was thrust in my face. I was left bereft and bewildered. I found out that for years I had been betrayed in every way a woman could be. We had never really had cross words or even disagreements—we were not close enough for that. I had had plenty to engage my attention while he gallivanted off on his various hobbies and, as I later found, conquests. Again, how sad. I politely asked him to leave, and after some equally polite resistance, he complied. That was that.

We divorced about seven or eight years after parting and, in the meantime, I had to earn my living. We were not rich at all; he had frittered away nearly everything we had on sheer hedonism. I was left in this huge house on my own with my two cats. But, that meant that I was never lonely and, in fact, I hardly noticed that he had gone. One welcome difference was not having to wash his shirts and cook his dinner every day. I certainly was not going to waste money on lawyers and I carried out the divorce proceedings myself with a little advice from my attorney niece. The whole thing cost a couple hundred pounds; I even asked him to reimburse me for that and he complied. That was such a painful time, but certainly not as bad as the very worst thing that happened to me. How did I survive that physical and mental agony?

The first disaster in my world was the sickening news that my beloved grandmother had been killed in a car crash. I just could not take it in and, for the first time in my life, I felt alone and frightened, shaken, and disoriented. Worse was to come within the year. My child was a spark of joy, certainly the sunshine in my life, and my reason for being. I lived for her—my blue-eyed, golden-haired daughter, Siân. For the first nine months, she was remarkably well and robust, and she was, of course, an exquisite child, as is everyone's. We were very close and I used to nurse her in a Welsh shawl, wrapped around my body, singing to her, with her slung next to my bosom. We both loved it. I still have the shawl.

They say that bad things come in threes, and they certainly did then. One morning, Siân was very fractious and we were both shocked when, at the minimal rebuke, she burst into violent floods of heartrending tears. When I took her to the doctor, he dismissed me as being a neurotic first-time mother. Even when it was obvious that something was really wrong, I was still treated as a nuisance. Not long afterward, her back was covered with a mass of tiny purple smudges, which I now know to be caused by bleeding under the skin. These were purpura. There was no internet to consult and we only had the GP for advice. This time he took me seriously and Siân was urgently referred to a consultant pediatrician. I cannot bear to dwell on the details of the months that followed but she was eventually diagnosed with Hodgkin lymphoma—essentially a blood cancer. The local consultants did not have the level of expertise to deal with this in such a young child and St. Thomas' Hospital in London became our second home for the next few months. Eventually, they realized a misdiagnosis and it turned out to be a very rare auto-immune disease called Letterer-Siwe disease. The next ten months were sheer and utter hell. Every time I went to her cot, I half expected her to be dead. She had too few red blood cells and I was able to give her my blood—anyone can have my blood because I am a universal donor (O rhesus negative). The horrendous medical and surgical interventions went on and on, and I realize now that the doctors just did not know what to do.

My only regret now is that she did not die more quickly. If she had, she would not have suffered so much, because suffer she certainly did. Even after all these years, I weep over her grave and think of her every day. What might she have been? Would I have had grandchildren? Would she have been more like me or her father? I never had any more children and do not know whether I would have been a good mother anyway. I always expect everyone to do their best, to excel at what they try. I know I am critical and would have probably been an unconventional mother. Perhaps Siân would have disliked me as much as I disliked my own mother—but I know I would always have put her first. I delighted in her existence. I desperately tried to protect her, but she ceased to be Siân on that cold January day. I stared in numb disbelief. I felt an electric shock and then the draining away of my insides. Not everything that left me ever came back. There is still an overwhelming void that nothing will fill or satisfy.

When the nurses gently led me away, I realized that the worst thing that could ever happen to me had happened. No disaster or misfortune would ever have the same impact, or ever hurt me as much. Since that day, I have been magically protected from being hurt by anyone. Nothing and no one would ever be as important to me as Siân. I am grateful that I had a daughter.

Being so protected can give one an air of arrogance. For the most part, I genuinely do not care what people think of me, and it has led to my being outspoken and direct. Nor am I afraid of anything or anyone. Perhaps that is why I have been able to cope with so many hideous and shocking sights and events during the course of my forensic career. I know too that my great concern for badly treated and neglected children and animals must, somehow, be linked to my daughter, her suffering, and her death. I know that some consider me to be fairly hard, but actually I am as soft as a marshmallow under the brittle exterior. Those who really know me are never frightened, but I know I can stop an undesirable in their tracks with one of my looks.

After I lost Siân, I was very thin and continued to suffer with my breathing. One day, in the medical school where I was working, standing over a bottle of urine collected from a ward, the professor commented on my cough.

"Oh, I have always had that."

"Well, I think you need to have it investigated."

And investigate they did.

In those days, they put you on a table that could rock in all sorts of directions, put a tube down your nose, poured a radio-opaque liquid into your lungs, swirled you around on the table until the lung was coated, and then took multiple X-rays. I remember watching the screen in fascination even though I felt I was drowning. The only way to get the white stuff out was to be turned over and thumped on the back until it was all coughed out. This procedure was called a bronchogram and has thankfully been replaced by the CT scan. No child or sick adult will have to experience this ever again. It was like being tortured by a group of white-coated aliens from a spaceship.

The diagnosis was swift. My right lung had completely collapsed into an abscess and my left one wasn't too good either. Within the month I was a patient at the same hospital where my daughter had died, in a ward right opposite Big Ben. Who would ever believe that, after a short time, one was not even aware of the bell booming out so regularly. Most of my right lung was removed and these were the days when anesthesia and pain relief were not particularly good. My memory of that time is acute; I was sucked down into a maelstrom of pain and misery, and the only escape was morphine. The euphoric sensation of being lifted up, up, and away from agony was truly magical. At first, I resisted because the needles hurt so much but, as soon as I asked for relief, it was refused. Obviously pain relief had to be weighed against the possibility of creating addiction. I can well understand how the next fix of a strong opiate can be the *raison d'être* for those subsumed by wretchedness and despair. It would be so easy to succumb when the reward for the prick of a needle is bliss and elation.

It took a long time, but I recovered and went back to work. My psyche and thin little body had taken too many beatings. But I am strong and robust mentally—if only I had the physical strength to match.

CHAPTER 12

Poisons

I may not have been able to go school as consistently as my classmates, but until I passed the entrance exam and ended up in that forbidding grammar school—that place at the top of the steep hill—I loved it, with the wonderful and kindly headmaster, Mr. Davies, and "Shorty" Jones, our teacher. What utterly gifted people they were. Mr. Jones brought the best out in all of us and yet managed never to show favoritism, nor make a child feel marginalized. I had not realized it myself, but more than sixty years later, at a reunion lunch, an old classmate told me that everyone knew I was the headmaster's favorite. She rolled her eyes and laughed when she told us all how he would always turn to me for answers, or ask me to set the class straight when we were floundering over some simple concept. I was full of facts imbibed from my encyclopedias and they were my favorite reading. After all, they were full of stories as well as being instructive and a great source of information and wisdom. My long hours spent in bed, when others were out enjoying games and adventuring, gave me the reward of knowledge, even if it was eclectic in the extreme.

My special friend was Jeannie Bruton and, when I was well enough, we would set off together looking for adventures. It was not far to the open hillsides outside the village where sheep and mountain ponies roamed free. We would be gone all afternoon, sustained by Marmite or jam sandwiches . . . if the ponies didn't get them first. They really were thugs at times, eager for more interesting tastes than the hillside turf offered. Jeannie and I were true companions. We picked bluebells and made scent in washed-out sauce bottles, ignoring the lingering sauce smell in the fermenting mush. We knew where to get frog spawn so that we could follow the progress of the tiny black commas in the jelly as they took form, and eventually hatched into tadpoles. By leaving bits of raw meat in the water, we even managed to get exquisite little froglets sometimes. We caught a slowworm and kept it in a big jar on our hall windowsill, although my mother eventually freed it from its glass prison and I bawled in fury that she had had the temerity to do such a thing without consultation. My dearest wish, which was eventually granted, was to have a penknife so that Jeannie and I could dig up pignuts. *Conopodium majus* is in the carrot family and it produces a tasty little tuber deep in the soil. We knew an area of grassland at the edge of a path where we could unearth these from a depth of only about four inches, which is fairly shallow for this plant. Anyone would think that we had unearthed treasure when we eventually got a few of these. We would wipe off the soil, scrape the outside a bit with our penknife, spit on them, rub the outsides on the hems of our dresses, and then eat them straightaway, even though they were sometimes a little gritty. They have a lovely, delicate taste and we always felt the digging was worth it.

Later, we would trek up the hillsides and pick wimberries. We loved them and came home with blue mouths, fingers, knees, and rear ends. Along with everyone else, we picked blackberries from September onward, and my knowledge of their leaves, picked up as a child, would later help me to tell the police how long a body had been in its grave. Simple childhood memories of being close to wildlife helped in a lot of my later casework too.

The big outdoors offers many treats that are quite safe to eat; the young leaves of hawthorn have a lovely taste and their berries later in the season have an interesting flavor and texture—although too many can be harmful. The berries of the elder bush have little taste, but they give stewed apple a lovely color when cooked with it. Local people called them gypsies' currants because, apparently, they were used as a substitute in cake-making. Most people know about the sloe or blackthorn, and I try to collect the tannin-rich, little plumlike fruit every year to make sloe gin, which is always delicious. If you try to eat a fresh sloe, your mouth dries up as the tannins in the fruit combine with the proteins in your saliva. It is a horrible sensation and you feel as though your tongue will never taste anything again. Freezing or heating the fruit, or soaking it in alcohol, destroys the cellular structure of the fruit and the tannins are released and lose their effectiveness. They certainly disappear in my sloe gin and, although I never manage to get the same flavor every year, it is always marvelous. Jeannie and I knew where and when to look for little treats and I am so sad that so many children today are too fixated by electronic games and texting to have adventures.

As mentioned, there are many wild plants that are safe to eat, but there are also many plants to be avoided at all costs. Plants and animals have evolved together over millions of years and, because no animal can produce its own food and must rely on plants at some level in the food web, if some defense against predation had not evolved, plants would have been chewed and sucked out of existence. Some plant defenses are exquisite and they show us how nature works toward balance. Protection can be mechanical, chemical, or a combination of both, and some involve mutualistic relationships with an animal.

The young leaves of the stinging nettle make good, nutritious soup—but to protect itself from too much predation, a combined mechanical and chemical defense has evolved. Specialized hairs on the leaves are intricately modified to inject a mixture of formic acid, histamine, acetylcholine, and serotonin into whatever touches it. This mixture causes a painful and burning rash, and there are records of dogs being poisoned due to running through extensive stands of nettle. Other plants, like wild parsnip and the giant hogweed, can cause a blistering rash if touched in sunlight. The books and papers written about harmful plants are legion and everyone should have at least a basic knowledge of them.

Defenses like this help to maximize a plant's survival in a competitive world. The thorns produced by many in the rose family are long, sharp daggers, and a hedge of hawthorn or firethorn acts like a wall of barbed wire; the thorns certainly deter foraging animals. Some *Acacia* species have thorns that become hollowed out by ants. The ant produces formic acid, which makes a sting long-lasting and painful, and it will protect the tree from browsing animals such as the giraffe, as well as other insect pests. There are many examples of such elegant symbioses in the woodlands, savannahs, and jungles of the world.

The number of plants with physical defenses like these, however, pales when compared to those that have evolved chemical defenses in their ongoing quest to survive and spread. The array of chemical compounds produced in the plant kingdom is bewildering, and many of them appear to be for protection—defenses against chewing insects and other animals, shields against parasites from entering their bodies, and some for which, as yet, a function has not been clarified. Plants, fungi, and bacteria are the astonishing factories of organic compounds on this planet. They can sustain the rest of the living world by providing food, but they also make compounds that cause discomfort, and even death. Vegans and vegetarians (nonhuman and human) are testament to plants' ability to sustain other living things, but there are plenty of examples of them being harmful or even deadly.

Throughout history, crimes have been committed using poison. It is held to be the preferred murder weapon of the female. We only have to think of the rumors and scandals surrounding Lucrezia Borgia, the illegitimate daughter of Pope Alexander VI, born in 1480; and Livia, wife of the first emperor of Rome, Augustus, who is supposed to have poisoned his figs in AD 14.

What is a poison? Does it differ from a toxin or a venom? All of these can be harmful to a person's well-being and can sometimes kill. Poison is a general word that covers any chemical that alters the normal body function. Inorganic substances like the element arsenic, or the compound potassium cyanide, can both be deadly poisons, but the ones delivered by animals such as snakes, scorpions, and spiders are termed "venoms" and they have a very complex chemistry. Most plant, bacterial, and fungal poisons are termed "toxins." Ironically, both venoms and toxins can be beneficial if taken in the right dose and will harm only if the concentration is too high. An example of this is the cardiac glycoside, digitalin, originally extracted from a species of foxglove (*Digitalis*). The genus produces a range of related, deadly cardiac and steroidal glycosides; in appropriate doses, digitalin (digoxin) can regulate heartbeat but in a large, unregulated dose, it results either in a lethal slowing down or speeding up of the heart, confusion, nausea, and even hallucination. It is easy to see how the humble foxglove could have been used to murder a victim in years gone by when medicine was relatively primitive, and before toxicology came into being. A victim might simply have been diagnosed with some kind of heart failure. Even the pretty pink- and white-flowered shrub oleander, seen all over the Mediterranean, has similar compounds and can cause cardiac arrest if even one leaf is eaten.

The toxins in foxglove and oleander can be found in any part of the plant but, in others, it may be concentrated in specific parts, or produced at certain times of the year. The red, fleshy part of the yew "berry" is not at all poisonous, but the black seed in the middle is deadly. The leaf stalk of rhubarb is delicious when cooked, but the leaf itself is full of oxalic acid, which is highly toxic. My mother used to boil the leaves in her old saucepans to get them gleaming.

Many plant toxins seem to have evolved as a defense against animals, especially the hordes and swarms of insects that crawl and skim around our planet. It is certainly one of the roles taken by terpenes. Urushiol, the truly terrible oil that caused the running sores on my legs after my brush with poison ivy when I visited the Body Farm, is a terpene, and it is present in poison oak too. The terpenoid toxin, thujone, is thought to be the mind-altering substance in the tissues of grand wormwood, an essential ingredient of absinthe, a much indulged-in drink by Parisians during La Belle Époque. Even though it can cause blindness and madness, it was certainly a favorite of Van Gogh, Gauguin, James Joyce, Toulouse-Lautrec, Picasso, Oscar Wilde, Proust, Edgar Allan Poe, Lord Byron, Ernest Hemingway and, most spectacular of all, Salvador Dalí. It makes one wonder if thujone played a part in the sheer inventiveness and expressiveness in the art; certainly some of the ideas and images produced by them seem quite mad to the rest of us. The output of some artists, whether visual, written, or musical, often seems to be beyond the realms of normality to me anyway. Many of us must harbor deep secret visions and thoughts that are suppressed by our rational selves. Throughout history, the creative intelligentsia have probably indulged in plant toxins to free their inner selves, and by doing so, give the abstemious, ordinary mortals among us a glimpse of their inner beings. Terpenes can be life-enhancing in other ways too. We are all familiar with Eucalyptus oil, camphor, turpentine, ginger, cinnamon, and the ubiquitous cannabinoids, which are toxins implicated in earlier dementia and schizophrenia. Even so, Cannabis is eagerly sought by many, and some of the more creative have admitted succumbing to its lure.

The King Kong of naturally produced poisons and toxins are the alkaloids. They are produced by bacteria, fungi, and plants, and are even present in some animals, such as poison frogs and toads, although these amphibians get the poison from the ants they eat and, presumably, the ants get it from the plant material they gather. The evolutionary advantage given to a frog from the plants is truly stunning.

Alkaloids are derived from the building blocks of proteins, and many have familiar names—atropine, nicotine, morphine, mescaline, adrenaline, ephedrine, and quinine. The vast array of alkaloids has thus been exploited for medicines; but the borderline between a poison and a therapeutic agent may, again, simply be concentration or dose. Tomato, potato, eggplant, pepper, and chili are all valuable foods, but they are all in the same family as deadly nightshade, with its tempting black berries, and *Datura* (jimson weed), with its beautiful white flowers. In reality, they all contain the alkaloid solanine and, although even some supermarket managers do not realize it, the greening of potato skin heralds the production of the poison. Solanine is a potent pesticide and it protects the plant from being attacked by pathogens but, so often, what is toxic to an insect, a worm, or a fungus will, inevitably, be poisonous to us too.

Strychnine is a well-known alkaloid toxin from trees and shrubs in the genus *Strychnos*, native particularly to southern India. A potent neurotoxin, it is well known because of its favor as the poison used in murder mysteries by authors such as Agatha Christie. It is rather an obvious poison for modern murders, but plants in the buttercup family—delphinium, peony, aquilegia—also accumulate deadly alkaloids in their tissues and, in 2009 in west London, Lakhvir Kaur Singh, after being abandoned for a younger woman, was convicted of spiking her partner's curry with monkshood (*Aconitum*), a beautiful purple flower from the same family. Every part of the monkshood plant is highly poisonous, and one of its effects is that it promotes motor weakness and paralysis, eventually interfering with the function of your heart and lungs. For this reason, like strychnine, it has a long history in murder, and after eating the curry, Singh's victim vomited, began to lose his vision and the use of his limbs, and died within an hour of arriving at hospital.

There are about two hundred species of *Aconitum*; they were used, variably, to eliminate criminals and enemies in ancient Rome, and to poison arrowheads in medieval warfare. Singh had laced the victim's curry with *Aconitum* and was eventually sentenced to life imprisonment for the premeditated attack. She was the first person to be prosecuted for murder with this type of toxin since the doctor George Henry Lamson killed his eighteen-year-old brother-in-law, Percy John, by delivering a poisoned slice of Dundee cake to his boarding school in 1882.

Fungi produce a most bewildering array of compounds, many of which are toxic, and many being helpful to us. Fungi are the source of many antibiotics, which are, of course, effective in killing bacteria. In nature, this prevents bacteria from swarming all over the fungal body, thus preventing it from absorbing food. Without the beneficial effects of fungi, nearly all animals and plants would fail to survive as they are essential for the ways they obtain their food; they are essential for the production of most of our food too. But, beware, the most poisonous have no remedy and many of these are in the fungal genus *Amanita*. The destroying angels (*Amanita virosa* in the UK and *Amanita bisporigera* in the US), and the death cap (*Amanita phalloides*), are particularly dangerous because they resemble edible fungi. However, Caesar's mushroom (*Amanita caesarea*) is safe and delicious and is sought after for its lovely flavor. In Europe and the US, *Amanita* is responsible for about 95 percent of all deaths in mushroom poisoning, with the death cap being responsible for about 50 percent on its own.

The problem with fungal poisoning is that the symptoms do not manifest themselves for at least six to fifteen hours, and in some species, up to a week or more, by which time it is too late even to attempt treatment. In any event, nothing can be done and death ensues. Certain fungi, then, seem the perfect agent for murder without detection. It is easy to imagine that, in the past, plants and fungi have been a rich source of poisons for nefarious purposes and details of the death throes of the emperor Claudius suggest that his fourth wife, Agrippina, was following in the family tradition by poisoning his food, this time with mushrooms, possibly *Clitocybe* or *Inocybe*, which have the highest levels of the alkaloid muscarine.

Many fungi produce alkaloids, the ergot fungus (*Claviceps purpurea*) being perhaps one of the most famous. It infects the female part of grass and cereal flowers, particularly those of rye, and produces dark bodies called "sclerotia" that replace the whole seed of the plant. The sclerotium behaves like a seed too and gets dispersed with the cereal grain, becoming mixed up with flour and finding its way into bread. The fungus produces a complex of alkaloid compounds; some have the effect of causing strong uterine contractions and this property was harnessed to cause abortions in the seventeenth and eighteenth centuries. Some of the alkaloids also cause mental confusion and vasoconstriction, especially of the hands and feet, creating a burning sensation—the so-called St. Anthony's fire. Where the sclerotia were used by doctors for the purposes of securing abortion, many women died and lost limbs because the dosage was so hit and miss.

The fungus also synthesizes lysergic acid, the precursor of LSD, and unfortunates who consumed the fungus, either accidentally or purposefully, were plagued by hallucinations. It has been suggested that the preposterous accusations made in the Salem Witch Trials in 1692 and 1693 were the result of people eating infected bread. Vast numbers of people all over Europe have died from ergotism, including forty thousand in Aquitaine in AD 944. The first known reference to this disease is from Assyrian tablets from about 600 BC and from India in about 350 BC. Wherever people grew wheat rather than rye, they were relatively safe as wheat is more resistant to the fungus.

Deliberate poisoning might be rare in the modern age, but the potential for intoxication is all around us. Swallowing a tiny amount of the alkaloid coniine, in the hemlock plant (*Conium maculatum*), will inhibit the working of your body's neuromuscular junctions, paralyzing you first in the legs and then spreading upward through the body until it reaches your lungs and finally kills you. Socrates must have had a nasty death.

One of the most potent plant poisons of all comes from the seeds of the castor oil plant, which contains high intensities of the toxin ricin. This is yet another class of toxin, a kind of protein that prevents other organisms from making their own proteins. Just four seeds from the castor oil plant can lead to an excruciating death from vomiting, diarrhea, and, in only a few days, multiple organ failure. In fact, ricin has such potency that it has been turned into a weapon of war, classified as a controlled substance by the 1972 Biological Weapons Convention, and used by various terror groups in assassination attempts against US politicians. In a famous case, the Bulgarian playwright Georgi Markov, who had defected to the West in 1968, was assassinated in Waterloo, London, by a member of the Bulgarian secret police who used a modified umbrella to fire a tiny pellet laced with ricin into his leg.

But although plant poison is not the modern murderer's most obvious weapon of choice, there are other circumstances in which a forensic ecologist might be called upon to look at plant and fungal remains where toxicological analysis fails. Certain toxins produce devastating effects on a body, but others are harnessed for other, mind-altering effects—and even taken recreationally. Plants and fungi have been esteemed for their psychotropic effects since prehistoric times.

The red fungus with the white spots (*Amanita muscaria*) produces ibotenic acid and muscimol, compounds with similar effects to lysergic acid diethylamide (LSD). Even reindeer seek out this prettiest of fungi, presumably because of the pleasurable experiences they get from it, and it has been used ritually by the peoples occupying the tundra as far back as anyone knows. One of the effects of these hallucinogenic compounds is that when under its influence, you might think you can fly, and many a death has needed investigation after devotees have thrown themselves from buildings with outstretched arms. Even Santa Claus, flying with his reindeers in his red-and-white suit, might be related to stories originally told by followers of the cult of this fungus. Many mushroom cults have existed, and still exist today, some simply associated with pleasurable experiences and others with religious ones. The basis of some religions, and even cultural behavior, seem to have stemmed from hallucinogenic experiences after consumption of some mushroom or other; and fungal emblems and depictions are common in ancient pictographs.

In popular television, forensic investigators are infallible and can always identify mysterious compounds central to the story, but it is chastening to realize that, even today, toxicology is often utterly impotent in identifying toxins unless the analyst is given some idea of what is suspected. Sophisticated analytical techniques are available, but just imagine the size of the library of reference samples needed to compare chemical structures of the thousands and thousands of unknown substances. If plant or fungal tissue or spores were involved, direct observation with the microscope can be successful in putting a name to the possible source of a toxin, but here an experienced botanist or mycologist is essential, and we are a dwindling breed.

Ancient cultures, and the rain forest native tribes persisting today, have possessed, and still have, intimate knowledge of the botanical and mycological heritage surrounding them. Many fungi and plants gathered from the steaming forests, and having psychotropic properties, continue to be part of their normal lives; they play important roles in tribal cohesion and structure. Many of the plants are toxic, but in the right concentrations and combinations that have been learned over eons of time by chosen elders, witch doctors, and shamans, ill effects seem to have been minimized. Brews of various combinations of tropical leaves, stems, and even roots can give heightened awareness and endow feelings of euphoria, well-being, and elicit religious experiences. One can well understand that certain native peoples have vast knowledge on safe dosages and the species mixtures needed for specific kinds of experiences to happen. But could you honestly imagine, say, a carpenter from Yorkshire, or a car salesman from Guildford, having sufficient competence to be a shaman? Well, surprising as it may seem, some in Britain (and probably elsewhere, too) supplement their income by ministering concoctions of mind-altering substances to groups of like-minded individuals.

One afternoon, I received a telephone call from a fairly excited police officer. He had taken into custody a man who had held a shamanistic ceremony and shared an exotic brew with sixteen people one summer's evening in 2008. Most of the group had been having a thoroughly predictable and enjoyable time, but one young man became agitated and then went berserk. None of the fifteen others, who also drank the same infusion, suffered any ill effects, only the hallucinatory and euphoric experiences they had been promised.

The brew the shaman had served up was called "ayahuasca" and it has been used, in various forms, for centuries—perhaps even millennia—by indigenous peoples of South America to induce healing, hallucination, and out-of-body experience. Ayahuasca brews are ordinarily based on the macerations of two rain forest climbing plants that, like so many in the rain forest, look rather similar even though they are not related. Invariably, the brew contains *Banisteriopsis caapi* and, depending on the effect the shaman is hoping his brew will provoke, various other plants such as *Chullachaqui caspi*. But, commonly, a partner for this mixture is *Psychotria viridis*, and this was the one chosen in this case.

Banisteriopsis caapi contains alkaloids that can inhibit the enzymes in our gut that would break down serotonin (the happy hormone). These inhibitors are called harmala alkaloids—harmine, harmaline, and tetrahydroharmine. Serotonin keeps us in a good mood; it contributes to our feeling of well-being, appetite, memory, and sleep, and, when we get clinical depression, it may be the over-efficient removal of this hormone by our own bodies that creates the problem. Some antidepressants act by preventing the naturally produced serotonin from being removed so that it is able to reach the active centers in the brain.

In ayahuasca, it is the second plant, *Psychotria viridis*, that provides the active psychoactive ingredient, dimethyltryptamine (DMT), and this is the compound that provides the out-of-body and other extraordinary experiences. Its compounds are kept safe from our body's destructive enzymes by the first plant in the brew.

The enzyme inhibitor must be present in any mixture or the active principle in other plants would not be able to survive the gut, get into the bloodstream, and pass the blood-brain barrier. Though traditional shamans likely had other, more fantastical explanations, it is the effect of DMT on the brain that is responsible for the otherworldly hallucinations; and the serotonin working so suddenly is largely responsible for the euphoria experienced by the participant.

The mind-altering substance DMT can be purchased, but it is a Class A psychedelic drug which is stronger than either LSD or the effects of magic mushrooms. This means that it is illegal to have it in one's possession, to give it away or sell it, and the maximum penalty for possession is seven years' custodial sentence. These were serious implications for this shaman operating in that leafy country town in the UK.

In that fateful summer of 2008, something had gone terribly wrong for the young man. The shaman, who had prepared the ayahuasca for him and his friends, had kept to the same recipe he had followed before. But this devotee's hallucinations seemed to be of a different order from those of the others; he ranted and raved and, when his friends took him away from the heart of the ceremony to calm him down, he slipped into a comatose state. He remained that way for four days and four nights. Throughout that time, his body still seemed to function—though raggedly, at best. Without being able to control the muscles, he became incontinent and his friends wrapped him in what nappies they could get together, keeping him as clean and comfortable as they could.

Indigenous societies may have seen vomiting and diarrhea as a necessary, and welcome, result of the ayahuasca ritual. Shamans would deliver their brews with the specific purpose of purging the mind of demons through hallucinatory experiences, but the concoction also purges the gut and rids it of intestinal worms taking up residence there. Indigenous societies, then, invoke purging and vomiting to keep people healthy but the young man who died in the care of his friends experienced no benefit from his part in the experience. These friends had no option but to call the police when he died.

The first time I had ever heard of ayahuasca was when I spoke to that police officer. Cases like these do not come up every day, and the novelty of having arrested a shaman had clearly impressed him. He had already sent off samples to be analyzed for the presence of DMT, and the results had come back positive. And yet the cause of death for this young man could not be established. The police knew that there was DMT in his body, but fifteen other people had drunk the same ayahuasca brew that evening, with no apparent ill effects. No doubt if the police subjected them to the same testing, DMT would have been found in their bodies too. The question we were confronted with was: Why him? Why had this young man perished where others had lived? Was something else involved in his very negative reaction?

The police had already made some headway in their investigations by the time I was asked to be involved with the case. When the group of depressed ceremony participants admitted that their friend also indulged in magic mushrooms from time to time, the police sorted through the man's rooms and found various flasks, a cookie tin, and a plastic container. A drawer from his bedroom contained a whole dried mushroom. What was the mushroom he had been keeping? What had been in these containers and flasks? Had it contributed, in some way, to his death?

I was tasked with providing answers to these questions.

My husband, David, was able to start piecing this puzzle together as soon as he saw the fungus; he identified it as *Psilocybe semilanceata* and confirmed this diagnosis by microscopic examination of its spores. This is the commonest magic mushroom and it actually pops up on our back lawn at home from time to time. It is one of the LBJs—the little brown jobs that are not easy for most people to identify precisely (although those who want to use them seem able).

I knew too that there were ways we could find out what had been in these containers. Working as an environmental archeologist, one of my most spectacular successes had been to prove that a "wine strainer," buried with a Druidic medical practitioner in his grave in Colchester, about two thousand years ago, had contained a concoction made from mugwort (*Artemisia vulgaris*), a common wasteland weed. Mugwort is a relation of wormwood and has long been used in herbalism. The wormwoods and mugworts contain vast arrays of alkaloids, monoterpenes, and many other compounds and, in the past, had sundry medicinal uses. People and domestic animals were dosed with mugwort tea to paralyze intestinal worms so that they could easily be passed out of the gut. Worm infestation was a fact of life for ancient people and clues to their misery have been gained from frequent findings of nematode worm eggs in ancient latrine deposits. As intimated from the field data and laboratory analysis, this Iron Age doctor in Colchester had been treating his patients with mugwort infusions for intestinal worm infestation, and possibly bacterial infections. Because of the plant's bitterness, he had been adding honey to the medicine. This was so convincingly clear because, although there were occasional pollen grains of cereals, grasses, or weeds—and although the strainer contained a huge mass of mugwort pollen—all the rest I found was from "bee plants." There is very little chance that the pollen of bee plants could have been included by accident because their pollen does not get into the air, or travel far from the parent flower. Honey, on the other hand, is rich in the pollen of plants to which bees are attracted because of their rich nectar.

To find out what the ayahuasca drinker man had been keeping in his containers I carried out the same routine as I had for the Druidic doctor's strainer. One empty flask had dried crusty material around its neck. I simply washed this off and processed it as I had the washings from all the other items. I was transfixed by the results as they started to reveal themselves under the microscope. One flask contained a dense concentration of Cannabis pollen and the pollen of mint; it looked as though an infusion of Cannabis and mint had been prepared as a drink. Another flask contained a very dense suspension of magic mushroom spores; again, it looked as though the fungus had been subjected to some preparation for consumption. The drawer was very interesting because, not only did it contain masses of Psilocybe spores, the pollen profile showed that the mushrooms had probably been collected from a grassy area near woodland.

It was obvious from my results that the young man was in the habit of not only attending ayahuasca ceremonies but also indulging in Cannabis and magic mushrooms. But did these, in any way, contribute to his death? So little is known about how the human body reacts when certain drugs are taken in combination. We know that the combined use of tobacco and Cannabis can induce psychoses in certain individuals, and we know that drinking alcohol can further aid the passage of DMT through the bloodstream and into the brain unhindered, but as to how the countless other drugs out there, whether natural or synthetic, respond in concert with one another, our knowledge is fragmentary.

Understanding the true cause of this man's death was paramount. The shaman had been arrested for possible manslaughter. His freedom hung in the balance. If there was no evidence of anything other than the active agent of ayahuasca in the man's gut, the police would be able to make stronger their case that the shaman alone was responsible for the dreadful outcome of the ceremony. If, on the other hand, the deceased had dosed himself with something else—something that had perhaps interacted with the ayahuasca, the shaman might be less to blame, and his liberty less at risk.

In cases like these, it does not pay to jump to conclusions. Mine are not moral nor legal questions. It must not matter to me or my findings whether the shaman might be prosecuted for murder, manslaughter, or some lesser offense. I must, at all times, restrict myself to the facts, to questions of who, what, where, and when: tangible questions demanding tangible answers. And these answers were to be found in the contents of the dead man's gut.

During a postmortem examination, it is routine for the pathologist to take samples of the gut contents. On most occasions, I have been somewhat bemused by the way this is done. Some pathologists simply put a kitchen ladle into the stomach and assume the ladleful represents the gut contents. To my mind, this is wholly inadequate. One can never consider the contents of a stomach to be homogeneous, with all the contents equally distributed throughout. By this method of collection, critical evidence could easily be left in the stomach. Sometimes, especially if the death is long after a meal, there will be little to sample anyway, but if the stomach is full, surely the contents warrant more comprehensive investigation.

In this particular postmortem, the stomach had certainly been sampled. I already had a mortuary jar full of opaque yellow liquid in front of me, with little floating globules of fat on the surface. It smelled strongly of orange, and this, at least, backed up the claim that the man had been sustained by orange juice during his coma before he died. This pale amber-colored liquid I was holding was only the beginning of the story, but as I studied it for the first time, I realized that it was unlikely to provide the information we needed. The fact that the man had continued to defecate, even four days after slipping into unconsciousness, meant that peristalsis—the wavelike muscle contractions that force food along the length of the digestive tract—must have continued throughout his inevitable decline. It was obvious to me that anything he had eaten or drunk at, or before, the ceremony would, by four days after the event, be in his lower gut; and if we wanted to know what else he might have taken, I needed to examine it.

"I'll need samples from the ileum," I said, meaning the stretch of ten feet of the small intestine that empties into the colon. "Could I please have a sample of the proximal and distal colon, and the rectum as well?"

It took a bit of a struggle to convince the pathologist that I needed the other samples of the gut but, eventually, I received just six—four from the ileum, one from the ascending colon, and one from the transverse colon. They would have to suffice. I did a standard preparation of these and eagerly examined them under the microscope.

There was little doubt that this person had not eaten for a long time, and perhaps fasting had been part of the shamanistic ceremony. A great deal of what had been in his gut had moved through his body and been collected by the nappies. The stomach contained little but orange juice, and there was nothing at all in the small intestine. It was not until I started looking at the colon that I had my eureka moment—there was plenty to report.

At first it was only a single orange pip; this was hardly surprising, given the amount of juice his friends had made him drink in their vain efforts to save his life. But then I found what I was expecting—a mass of magic mushroom spores and a good complement of *Cannabis* pollen. They were more abundant in the transverse colon than the ascending because they were closer to the rectum. The man, it seemed, had been taking mind-altering DMT in the ayahuasca, and mind-altering cannabinoids in the *Cannabis*. He had also been drinking the magic mushroom "tea" from the flask that would have contained psilocin; in the gut, this breaks down to psilocin, which is the hallucinogenic component in the magic mushrooms. What a cocktail! This seemed heady enough, but what I saw next added an extra layer to this dead man's experiences. I could hardly believe the number of seeds. Not just any seeds—these were from the opium poppy (*Papaver somniferum).*

Bakers use opium poppy seeds to coat bread and rolls, and perhaps you remember a popular investigative TV program where they tested Angela Rippon, a well-known journalist and TV presenter, after a claim that a power station worker had been sacked because, in a routine test, opiates had been found in his blood. He claimed innocence and said that he had been eating seed-coated bread. The program wanted to test the hypothesis that bread rolls could, indeed, yield measureable opiate in the blood. This was confirmed by the analysis of Angela Rippon's blood after eating such bread for just three days. The seed contains only a tiny amount of opiate but, of course, the more you consume, the greater amount you will absorb. In reality, the amount of opiate in the seed depends on the plant, its growing conditions, and harvesting methods—but there is little doubt that, if you eat the seeds, you might have opiate floating around in your body. The seed normally contains between 0.5 and 10 micrograms of morphine per gram, whereas a medically prescribed dose will be anywhere between 5,000 and 30,000 micrograms. Nevertheless, seeded bread is a problem for the World Anti-Doping Agency–accredited laboratories and a sports competitor may test positive for morphine if the level in the urine is greater than 1.3 micrograms per milliliter. It has become such an issue in the workplace that, in the US, authorities have increased the tolerance level to 2 micrograms per milliliter.

There were so many seeds in the dead man's colon that the story seemed to write itself in front of my eyes. I could visualize, now, the way it might have been; how, sometime before the ceremony, the man had brewed and drunk his own mixture of Cannabis and mint tea, and then magic mushroom suspension. Not yet satisfied, he had then eaten a mass of poppy seeds before setting out to the shaman's ceremony to enjoy his ayahuasca. He had not known it, as he put the cup to his lips, but already in his body were toxins in such quantities and combinations that the taste of that ayahuasca would be the very last one of his waking life.

Some forensic toxicology colleagues from northern Europe have since told me that the practice of eating masses of poppy seeds is growing in Europe. Perhaps I have too busy a life to ever contemplate, or even desire, the kind of out-of-body experience that is clamored for by certain others. Perhaps I would be too frightened anyway. But my mind boggles at the lengths some people will go to in order to experience a state of mind impossible without chemical inducement. I find it intriguing and inexplicable that they willingly indulge in so many toxic substances, including commonly consumed alkaloids—caffeine, nicotine, cocaine, heroin, and morphine—to lift themselves out of the ordinary and mundane.

Even I, though, occasionally have a couple codeine pills to kill a headache.

DMT, it is said, takes you rapidly into an intense alternative reality. It can take a long time to recover from the experience and, if you have any mental problems, it can exacerbate the symptoms. Some who have indulged say that they keep having flashbacks to the experience, and there is little doubt that this powerful substance alters the mind, sometimes for long periods. Some say they have been shown the face of God. Some say they have visited alien worlds and spoken to alien beings. A common experience is of being able to speak to owls in a strange language, while yet others have reported "going to hell."

Punishment after death, for evil committed during a lifetime, is a common concept in religion, and artists throughout history have depicted hell as a place of eternal agony. The fantastical conjurings of Hieronymus Bosch and Brueghel depict a view of chaos that might well be hell. Whether their imaginings emerged under the influence of some kind of chemical agent, no one could say. But, some of the equally bizarre works of the later surrealist artists may have been inspired by absinthe and/or other mind-changing substances. No rational person could take seriously their concepts of hell and damnation as being meaningful. My own view is that everyone's hell is personal and consists of their own discomforting horrors. The man who died after seeking excitement, uplift, or fantasy at the ayahuasca ceremony might have been in his very own hell during his agitation and raving. What hallucinations did he see whirling around him as his friends struggled to restrain him and make him secure? His screams and frantic movements might suggest that he was suffering, and no amount of care from his friends could negate the effects of the various plant and fungal toxins doing battle in his brain. We are, after all, but chemistry.

The DMT from the ayahuasca brew might very well have been a contributory substance in the lethal mixture this young man had consumed, but because his body had yielded so many different kinds of pollen, spores, and seeds from hallucinatory plants, it would have been difficult to lay the blame for his death purely at the feet of the shaman. In court, he was deemed not culpable of the crime for which he had been charged, and he was, instead, convicted of possessing a Class A drug, quite separate from the ayahuasca ceremony. He was given a short custodial sentence but it was not long before he was out, and the last I heard he was still holding his ceremonies, still mixing his hallucinogenic brews— still opening the doorway to "other worlds" for those who came to him, eager for a taste.

It seems sad to me that people are desperate to leave their own reality for another. Perhaps some think that this world that we have—with all its natural beauties, the things we can and cannot see—is not enough for them. As for me, I know I will not be going to a place called hell, virtual or otherwise. Death is coming for all of us, for you and me, and everybody out there. Better, first, get a life.

CHAPTER 13

Traces

It was after dinner on New Year's Day and, abandoning my resolution to give up wine, I settled down with a glass, Mickey on my lap, and a list of TV programs to watch that evening. Then the telephone rang in the hall.

"Hello, Pattie, Happy New Year! We have a job for you; could you be at headquarters by six tomorrow morning?"

It was Dougie Bain, the senior CSI at Hertfordshire Constabulary.

"Will you contact the experts and put a team together? I need you here by six a.m. tomorrow morning—skeleton found in woodland."

Same old story, the inevitable dog had found a murder victim while walking with its owner. It often happens to the hapless who are up and about before the world stirs out of bed. Choosing the right people for the job was easy and, thankfully, quick phone calls established that all were available.

The following morning there were few drivers on the M25 and I got to police headquarters in Welwyn Garden City within an hour. Trying to look bright and enthusiastic, though feeling light-headed from a lack of sleep, I absorbed the details in the short briefing before we set off in convoy to the deposition site.

"I've followed your teaching, Pat. You go in first and the archeologists after the entomologist. The other forensic scientists and the pathologist can do their work after that."

This was a perfect strategy for getting the maximum amount of evidential material from plants, soils, animals, fungi, disturbed vegetation, and footmarks. I could make sure that all classes of potential evidence would be preserved; every specialist would be able to retrieve the maximum information possible. We kicked and crunched our way through the russet leaves carpeting the woodland floor, then slid down a steep bank. A tiny stream meandered along the bottom of a little valley and, at one point, split into two even smaller ones, each flowing around a little island, merging into one again on the other side. The exceptionally neat grave on the center of the island was distinctive, and it was obvious to those of us who work with buried remains that the offenders had chosen the site carefully so that it could be found again easily. As I have previously mentioned, often murderers cannot resist revisiting the place where they have left the victim, possibly to check, or even to gloat. A log of wood might be put over the grave as a marker, or it may be near a distinctive landmark. Here, the little island marked the spot.

I left all the others and crossed the stream to the grave. The sight that hit me is still vivid in my memory. Defleshed feet poked out of what looked like a shallow mud bath, and the empty eye sockets of a grinning skull were staring at the muddy puddle where its belly should have been. Maggots and other scavengers had done a good job of picking the bones clean, and I was left alone to observe this unfortunate, and its surroundings, with everyone else well away on the other side of the stream, stamping around to keep warm, smoking, laughing, and eating chocolate.

The Forensic Science Service staff and pathologist have traditionally considered their work to have primacy at crime scenes, but I had schooled these police officers in what could be lost if they did not modify the usual protocols. After all, it did not need a forensic pathologist to confirm that the owner of a skull, picked clean by birds and rats, was dead. It might be a statutory requirement for the pathologist to declare that death had occurred, but he certainly did not need to be first at the graveside—not if it meant disturbing the natural environment that might yield vital clues.

It was cold and I had put on too many layers which, together with the all-enveloping Tyvek suit, meant it was difficult to move easily. My nose and eyes kept running because of the cold air, my fingers were stiff inside the protective gloves, and my handkerchief was out of reach; I was uncomfortable.

The first question everyone wants answered is: How long has the corpse been here? Vegetation in a situation like this often gives the first clue. Digging the grave had disturbed a stand of bracken fern that had been growing all over the island, and it was important to examine it carefully. A little distance away from the grave, I dug away the soil around the cut ferns until I reached the underground stems. These had been roughly cut but they still had their dormant buds, which would grow to replace the damaged leaves. Yes, the buds were there and, compared with others along the stem, they had become swollen and extended.

There were also tiny fragments of green leaf in the soil. Chlorophyll is a fascinating molecule. It is resistant to breaking down when prematurely detached from a living plant, and takes a long time to decompose; fresh leaves can remain green in soil for months, long after the intact plant has gone brown and died. This was the situation here. The stimulation of the buds on the underground stem, and the green leaf fragments in the soil, suggested to me that damage had occurred at some time in the previous summer—certainly before early autumn when these fern fronds start to turn yellow.

I staggered up from my knees and called over to the waiting crowd: "I'm pretty sure this is a late summer grave." A mock cheer went up. One of the most important pieces of intelligence is the timing and dating of events. It helps investigators to check the alibis of suspects against the calendar.

My next move was to collect comparator samples from around the grave, from the grave-fill soil itself, and along any likely pathways that the offender had used. With a little luck, I would later be able to compare the profiles of these areas with any footwear, vehicles, and implements retrieved from possible suspects. Considering that a single fern frond can produce 30 million spores in one season, I expected this plant to be an important contributor to the jigsaw of evidence retrieved from this site.

As I prepared to sample the grave, I put on my face mask. I did not want to risk coughing, sneezing, or even breathing DNA onto the corpse, or any of the hairs on my head getting into the grave. The forensic practice of "suiting and booting" not only protects the crime scene from human contamination, it also guards against anything nasty being picked up from the corpse. I was co-cooned by my suit, cut off from everything and everyone. I had scraped up my samples and put each in a separate polythene bag, made a list of all the plant species I could find, took pictures of anything I thought was relevant, and made sure that all the samples were labeled and logged accurately. It seemed to take an age because of fumbling cold fingers encased in vinyl.

"Okay, chaps, come over," I shouted to Peter Murphy and Luke Barber.

Peter was an environmental archeologist like me, but based at the University of East Anglia, whereas Luke worked in the Sussex Archeology Unit, which was administered by UCL. We crouched down together and they quickly removed much of the overlying material to reveal a great surprise to all of us.

As soon as they scraped away the mud slurry from the surface and exposed a more compacted soil underneath, I gasped.

"I'm wrong—this is not a summer-dug grave!"

I pulled a face as I realized that the grave had been refilled in the autumn or winter. This was obvious because the grave-fill contained brown leaf litter that had been carpeting the ground in the not-too-distant past, and just as it was now on the other side of the stream, being aimlessly kicked about by other team members. A summer soil might have contained bits of green leaf and debris, but definitely not such a quantity of brown, shriveled leaf debris. How could I have been so wrong? But was the grave filled during this winter or last winter? The victim looked as though he had been in the ground for a very long time. The leaves might well have remained for a year, but there was still the puzzle of the bracken stems. I hunched down and kept scrutinizing as the entombing deposits were expertly removed by precise and rapid troweling.

Archeologists and biologists view soils differently. As they dig, archeologists describe variations in soil color and texture in terms of "contexts"; each one possibly representing an event or episode that recorded human activity directly, or indirectly. A biologist, though, will regard a soil profile in terms of horizons formed by natural processes. Both are valid concepts which help us each form our own conclusions about disturbance and development. In a recently dug grave, the backfill is just a mishmash, but we can glean evidence both from what is included in the fill and its dissimilarity to the underlying "natural." Archeologists and scientists are pre-adapted for forensic work; it is second nature for every item and action to be meticulously timed, positioned, measured, and recorded. However, one of the most unenviable jobs is deciphering scrappy notes from soggy, mud-spattered notebooks, where what seemed perfectly clear at the graveside can be unintelligible back in the office.

We all gasped as more of the body became exposed. The skull and feet protruding out of the mud bath had been picked clean by maggots, birds, and rodents, giving the impression that the victim, in this strangely exposed burial site, had been dead for a long time; and, since there was virtually no evidence of interest by larger scavengers, we all thought that we might be retrieving a complete skeleton.

"Good grief!" exclaimed Peter.

He was not prepared for the slimy, pallid, well-preserved flesh under his trowel. He was used to seeing only ancient bone, which usually has about the same impact on the excavator as any other artifact. Luke was more relaxed because of his experience in excavating war graves. Our eyes met over our masks in sympathy for poor Peter, who was gagging as the stench grew in intensity.

I had already collected samples of the mixed grave soil and of surface samples all around the grave edge. No one disposing of this body could have avoided picking up this soil—soil that would have plenty of condemning trace evidence. After a couple hours of intense activity, the whole body was exposed and the rest of the forensic team could enter the inner cordon to collect their own classes of evidence. At last, the pathologist was able to declare the victim dead.

The undertakers arrived. It may surprise you that they often attend crime scenes, whatever the location, to carry a victim's remains back to the mortuary. To see three men in formal black suits and ties, white shirts and shiny shoes, slip-sliding down the bank, trying to keep their dignity and respectful demeanor is an incongruous sight.

The body was bagged and carried away by the silent men in black, scrambling and stumbling up the steep bank. Body gone, most of the police and others on the other side of the stream could be relieved from duty, and some were glad to be racing off to the warmth of the mortuary.

In the old days, the police would dig out the body as best they could, undertakers would take it away, forensic teams would take samples for various purposes, and that would be that. Not now. The crime scene is a precious place for clues to offenders; it must be combed and searched very thoroughly. And archeologists must be employed to retrieve human remains in the most meticulously recorded way possible. Archeologists always insist on revealing the original cut of the grave and, now that the grave was empty, Peter and Luke kept troweling in an attempt to find it. The floor of a grave can preserve footprints and tool marks made by the digger, and these can be measured and compared with those of an eventual suspect.

The unremitting thoroughness of the diggers had revealed something very damning. The soil that had been under the body contained no leaf litter in it at all, even though the bottom of the grave had still not been reached. This could only mean one thing. Realization then dawned. I had not been wrong at all—the grave had certainly been dug in the summertime, but it had been re-filled. Sometime later, when the leaves had fallen and were covering the ground, it had been dug again, but this time not to the bottom of the original grave. The victim had been put in the newly excavated hole, but the soil used to fill it was mixed with the current season's leaf fall. This murder was premeditated; the first grave had been dug in anticipation of the victim's death.

After everything had been recorded and photographed by police staff, and we were happy with our samples and notes, we jumped over the stream and climbed the slippery bank, panting and sweating in spite of the cold air. Luke went back to Sussex while Peter and I carried on to the mortuary, where work was already under way. As usual, I donned scrubs and readied my equipment, but there was little for me to do. The corpse was on its front—what a sight. This was not a skeleton but an almost per-fectly preserved man with his hands tied behind his back. The washing away of the grave soil from his front by rain had en-couraged scavengers to pick him clean wherever they could reach, but that was only his face, extremities, and chest. The cold and wet had preserved the rest of him beautifully, so much so that a perfect set of fingerprints was obtained from his hands. We had started very early in the morning, but by 3:00 p.m. there had been a pos-itive identification from NAFIS—the National Automated Fin-gerprint Identification Service. If a match is present in the database, the information can come back within about fifteen minutes. Here, there was a match—he was a known offender and his prints had been recorded in the database. He was a young, perfectly legal Albanian immigrant and had lived in London.

What had killed him was very obvious—a deep knife wound in the chest. Intensive work by the constabulary quickly identified his home, his associates, and what had prompted his murder. We were shown a picture of him—what a handsome young man—and I reflected on how such good looks and vibrancy could be changed so soon after death. He had been laundering money for his illegal Albanian associates, but temptation had overcome him, as it often does. Instead of making sure that the funds had reached their families back home, he had frittered it away, having a good time. Retribution had been planned, rather than been doled out in anger. The grave site was carefully chosen so that it would be easy to find again. It was on a little island in the stream gurgling through that tiny valley in that beautiful woodland. It must have seemed remote and safe to the perpetrators, but they had reckoned without the intrepid British dog walker.

I shuffled off to the changing room in my too-large boots, got back into my warm tracksuit bottoms and tennis shoes, and I drove back home with country-and-western music blaring all the way, drowning out thoughts of that dismal, long day. I had left in the dark before dawn and was getting home in the blackness of that late winter night. Mickey was waiting for me and we both had supper—me beans on toast and Mickey some poached trout. I wondered idly if I would be able to contribute anything other than my field report and interpretation of the grave stratigraphy, but I did not have to wait long to find out.

A few days later, the phone rang.

"Pattie, we know his car is back in Albania; we know it was

used to take the body to the woodland and we've seized lots of footwear and clothing from the gang for you to examine."

"Great, Dougie. I'll start on the footwear for comparison with the woodland as soon as I can."

"Yes, and I want you to have a go at the car too."

"Don't be daft, Dougie. It has been driven across Europe and the driver must have been in and out of it all the time."

"Well, we want to try."

I thought this was being unrealistic. I had shown the power of palynological trace evidence many times, but this was a bit much. I did not have the energy to argue and, shortly afterward, Peter Lamb, principal scientist at the Forensic Science Service, and I were winging it to Milan Malpensa airport in Italy, where we could get a connection to Tirana, the capital of Albania.

We were greeted by a British police officer who was a member of the investigating team and who had been out there for some time; he had certainly become used to the luxury of the hotel services and knew his way around. There was a lot of jovial bantering when everyone met up, but Peter and I were tired and, after a very good dinner, went off to our rooms early.

Next day, meeting the Albanian police officers who "were at our disposal" presented quite an eye-opener. These men all seemed huge and no one spoke any English except an assigned translator whom I disliked at first meeting. He seemed shifty and we soon realized that he was capable of bending the truth to suit his convenience. At first they all ignored me completely until Dougie introduced me as the person who would be in charge of operations for the environmental work. I can still remember their faces—they really did look stupefied that this diminutive female was capable of directing anything, let alone operational work.

Peter had to examine the car in the hope of finding fibers, blood, DNA, or any other trace evidence that might match those in the samples he had obtained from gang members' clothing and, of course, of the victim. On the other hand, I needed soil samples to eliminate those places frequented by the gang from those in the Hertfordshire woodland. The first place to start would be the main suspect's London home, and then the other addresses occupied by the rest of the ring. Apparently, the police had found the footwear and clothing all over the place—bedrooms, sitting rooms, and even sheds—a shoe here, a jacket there—and they just collected anything they thought might be relevant. One thing that makes life difficult for the analyst in this kind of situation is that associates often share clothes and footwear. One way to establish ownership of any of the items was to do DNA analysis. This certainly works on the inside of footwear, and one can even determine mixed profiles so that more than one wearer could be identified. Then, if the trace evidence on that item matched the burial site, further work could go from there. The residency of the gang was as complicated as the cache of clothing and shoes. Who was staying where and when? It is always difficult when dealing with criminal illegal immigrants; they learn how to cover their tracks. We could have been working on a huge mass of items for a long time, so the police decided to apply Occam's razor and they focused on the car.

I will never forget entering that Albanian police compound— the prison had the same entrance for security. It was like something out of a Dracula horror film; the gate was huge and had big studs. It opened into a wide yard with piles of rubbish along one side and some vehicles parked in makeshift garages along the wall on the same side as the gate. The red car was there all right. Before we could do anything, social niceties had to be observed. I was ushered into an office alongside the ramshackle garage, and the room seemed to be full of huge men, all bending down to smile at me, showing off white and gold teeth and overpowering me with garlic fumes. They definitely looked foreign—all dark and heavy stubble, but very gracious. They insisted on showing me their pride and joy—their fingerprint system. Rather than the system we use in the West, they employed one used by the Russians and even some of the British officers thought it was superior to ours. I could not judge, but it was fun to have my prints taken so that they could show me the sequence of events from ink to image.

The result that my fingerprints were so faint surprised them; you could hardly see them at all. Through the translator, they told me that I obviously did too much housework and had scrubbed them away. I suppose my passions for cleaning and bleach had a lot to do with the disappearance of my ridges and furrows that distinguish one person from another. Although the population of the whole world has not been tested, there have never been two sets of fingerprints ever found to be identical, and even those of twins, derived from the same egg, are different. Apparently, genetics play a role but fingerprint formation is more to do with a basal layer of cells in the fetal skin being sandwiched between the dermis and the epidermis, which wrinkles as pressure is exerted by growing, underlying tissues. The way fingers are used in life, and the multitude of little scars that are inflicted on the skin, also play a part in the distinctiveness of any print.

The Albanian police rarely had the opportunity to meet Westerners and they were keen to show us examples of their work. As we trundled across the yard toward the main building, I looked up at the faces behind, and arms poking out of, the barred windows, and wondered why so many of the prisoners were looking down at the yard. A bell clanged, the gates opened, and in flooded a mass of drably dressed women and the occasional man, carrying laden baskets covered with cloths. They all hurried across the yard and disappeared into the building.

"What's going on?" I asked the translator.

"It's feeding time."

"What?"

"The families are bringing their food."

"But what if they don't have anyone?"

He just shrugged and walked toward the main door of the building. What a contrast between the treatment of our prisoners and theirs.

On the other side of the yard, we were proudly ushered into a large room without windows. It was lined with long tables and the biggest array of guns I had ever seen. They all looked vicious and frightening, and had all been seized during active duty against criminal activity; they brought the possibility of death too close for comfort. I politely listened while a tall, thin man with a stoop gave us a detailed description of each weapon, with technical information that mostly went in one ear and out the other as the stilted translation was delivered. I smiled and nodded repeatedly like an automaton and, out of politeness, asked an occasional question via the translator. The only gun name I recognized was Kalashnikov, and they had lots of those. They made me shudder as I imagined them being used, as they inevitably had been. They also reminded me of a weak joke I had heard somewhere: "The English language is like a Kalashnikov. It will take you anywhere." I suppose that is true, though, up to a point.

The tables of guns also took my thoughts to my paternal grandfather, a man who rarely spoke unless it was to my lovely

Welsh, roly-poly grandmother. I do not think I ever exchanged more than a few words of polite greeting with him, and I found him intriguing, even though my cousins thought he was just grumpy. No one was allowed to make a sound if the news was on the wireless, and we crept around like little mice, terrified of rebuke. No one wept at his funeral and he was hardly missed. I later found out that he had been a machine gunner in the First World War and had experienced a bad time. I suppose he must have used the antiquated equivalent of a Kalashnikov and killed a lot of men. Poor soul—he must have had deep thoughts at times. I wish I could have the opportunity to talk to him now. All we know is that he was an actor, and was in the first moving picture ever to be shown in Britain. This was, of course, in the time of silent films, and he used to tour with a little film company that held film shows in tents, all up and down the Welsh Valleys. He was locally famous as Dick Turpin. At the end of each showing of *Dick Turpin's Ride to York*, he made the grand finale—fully masked with cracking pistols, riding onto the stage on a horse. He captured the heart of Gladys Blodwen, married her, and became a grocer.

We were taken to the document analysis department and, although I was reluctant because by this time I was hungry and my back was aching, what they revealed was utterly fascinating. They demonstrated the inventiveness and ingeniousness of the fraudsters in producing fake documents. These cheats were artists and supreme technicians, and might have been miniaturists in previous times. But the forensic investigators were usually a step ahead. They showed us how to detect alterations of numbers, letters, and pictures under their microscope. I was greatly impressed.

I was desperate to get to work but Peter Lamb was still meticulously examining every nook and cranny of the vehicle. I later learned that, even with weeks of work back in the UK laboratories, he failed to find any link between the deceased and the reference samples and car. As we waited for Peter to finish, the mild hunger pangs got worse and, although lunch had been promised, it failed to turn up, so one of the men went out to buy pizza. When it arrived, I sat on a wooden box in the yard about to start eating when a very thin little cat approached, followed by a tiny kitten and then another. Oh God, this was what I was dreading. I picked cheese off my pizza and the little cat gratefully gulped it down immediately. I gave her another, then some of the bread. I ended up licking off all the tomato so that it would not contaminate it for her and, very soon, she and her kittens had polished off nearly all my lunch. I was left with some tomato-stained crust. The Albanians looked at me in utter amazement and I rather got the impression that animals were not high on the Albanian priority list. This made me miserable and I have never forgotten that little cat.

Eventually I was able to get at the car. It would have been just too complicated to sample the footwells adequately in the yard, so I asked the exhibits officer to bag all the mats and record them as exhibits. I also removed the rubber covers from the clutch, accelerator, and brake pedals. I brushed the footwells with my small stiff brush, using a new one every time I sampled a different surface, and put the sweepings into labeled plastic bags. I examined the whole of the interior, including the trunk, taking samples wherever I thought there might have been an opportunity for trace evidence to be deposited. I would normally examine the chassis, but there was little point in doing that here because, of course, the vehicle had not been anywhere near the crime scene. It had simply stayed parked on a road while the victim had been marched to his death on that little island, deep in the woodland, back in the southeast of England.

Gleaning as much as I could out of the car had taken all afternoon. The police officers had been milling around in clouds of cigarette smoke, obviously bored, willing me to get on with it and get out of there. Eventually, there were smiles all around when I announced I was ready to go. The next major hurdle was to overcome the resistance of the local police to our meticulous way of doing things. Our officers had informed them that I needed to go to the main suspect's family home in Albania to collect samples so that I could separate that place from the original woodland grave site. There are nearly always some overlaps in the pollen and spore profiles and it is pivotally important to know what may have contributed to the results obtained from the various exhibits.

In a country like Albania, where the world of forensic detection is very different from our own, convincing local police of the importance of our methods proved to be an unnerving struggle. They simply refused the request to take me to the remote village where the suspect had lived. This was where he had fled after the murder had become news in the UK. The victim's family lived just a few hundred meters from that of the suspect, and now a deadly feud had developed between them. Everyone knew everything in those remote places.

On and on, their protests fell like hailstones:

"It was miles away; the roads were bad; there was no need to do this; we don't even believe you can make links with the grave; it is impossible—you are being stupid; it would take all day." "We will take her on a trip around the town a few times so that she will lose her bearings—she won't know the difference."

Dougie just grinned. "Oh yes she will."

So a convoy of badly maintained, scruffy vehicles set off the next morning to this feudal village in the hinterland.

I still do not know where they took me, but I do know that the journey made me lose the will to live. As this strange convoy bumped and swayed up the center of the dirt-track road through the village, with billowing clouds of buff-colored dust in our wake, heads were turned from the fields, eyes peeped from windows, and we were watched intently from doorways. We eventually arrived at the main suspect's family home. By this time, he was incarcerated in the prison we had left the previous evening. I wondered about him at mealtimes because this village was a long way from Tirana and the village where he was born, and I doubted that he would get many visitors. How would he get fed? This thought did not engage me for long as I was pleasantly surprised when we arrived at the neat, well-presented little farmhouse, with a yard bordered by vines, and an abundance of vegetables in the plots out front. People appeared as if from nowhere.

Several women of various ages, all looking rather weather-beaten and work-worn, silently edged forward accompanied by a flock of olive-skinned children. All were led by an older man, possibly in his late fifties. Everyone was dressed adequately in their daily work clothes, the man in faded cotton trousers, an old polo shirt, and a loose leather waistcoat, the women in long, drab cotton skirts, overtopped by loose faded blouses; they wore headscarves that just covered the back of the heads and their long hair. This was a Muslim country where one might expect women to be all but hidden but, presumably, because of their hard work in the fields, having to wear something that covered the full head was just impracticable. The children seemed to be dressed in any old thing, obviously mostly mismatched hand-me-downs, with no hint of coordination, or of preference. I contrasted these quiet little boys with the fashion-conscious, demanding ones I knew back home. The difference was stark. This family lived by subsistence, their grimy nails a testament to their hardworking lives. I could now understand the desperation felt by the illegal immigrants when the money they so wanted to send back home had been squandered by their compatriot.

The family was obviously in deep distress at the eldest brother being in prison for the murder of a neighbor's son. They all crowded around us and the father suggested that we move into the main room so everything could be explained. Why were we here? What had this to do with them? The son, brother, and uncle of the family was in prison, but "he had done nothing wrong." There was an uncomfortable hiatus in proceedings when the head of the household was told that I wanted to ask some questions. I knew by the face and body language directed toward me that I was the crux of the problem. No woman was allowed to sit with men for discussion in the main room. But, without me, no one knew what was needed; I was the only one who knew what questions needed answering. The situation was explained to the family and they all agreed that I should be bestowed with honorary manhood while I was their guest. Honorary manhood . . . I sighed inside.

Immediately past the threshold, the house was as clean as a hospital, and every surface gleamed, even the floor. The table in front of us was covered with an old but beautiful lace embroidered cloth; cups appeared and, not long after, a pot of coffee. I was surprised that they also produced some tea for me. The British and

Albanian officers, the father and I sat around in this special little room while the women and children all crowded into the doorway, not daring to pass the threshold.

Straightaway, there were exchanges between one Albanian officer, the translator, and the father. We had no idea what was being said but there was plenty of animated gesticulation and anger seemed to be rising. I cut across the translator's diatribe. The men's heads jerked around. The sound of a woman's voice was obviously startling.

"I am only here to find the truth. If your son is innocent, then I will be able to say that he must not be convicted. I am not on the side of the police. I am neutral."

That, of course, was absolutely true and I have often thought it was only my sincerity that helped pacify everyone that day. The father then looked at me and what I had said obviously hit a chord; he said he would help us "so that we could find his son innocent." I had a lump in the pit of my stomach that lay there for the rest of the day.

I explained that I needed to take some samples of the soil in the garden to compare with the soil from the crime scene. They were so confident of their son's innocence that they helped me as much as they could. While I crouched along the well-worn paths in the garden, the children followed my every move and took it in turns to approach me, smiling, each with a flower. Again, I felt the lump, but this time it had reached my throat. They were such lovely people—very simple and hardworking. This family was different from the people I had met in the city and at the prison. I very rarely interact with any family involved in casework. There would be a danger of favoring a suspect because of sympathy, or even empathy, and these could lead to cognitive bias. Subconsciously, I might want him to be innocent, and this must be avoided at all costs. It was hard but impartiality is imperative.

When I had sufficient samples to represent that place, I asked about the whereabouts of the nearest woodland. The suspect's father waved toward a ridge of steep hills on the horizon. My gaze followed the sweep of his arm. They seemed forever away. But it was imperative that I make a list of species in any woodland within striking distance of the house. From what I had already been told, it would be impossible to get a species list from the university, and my previous attempts at searching the internet for information about the country's botany, and distribution of its plant communities, had proved futile.

There was nothing else for it, we had to get there so that I could assess the environment myself. After exchanging pleasantries with the troubled and anxious father, our scruffy, dust-covered convoy set off toward the mountains on the horizon. The drive seemed interminable and my eyes were constantly scanning the vegetation. The mountains were delightful and I would have liked to spend a holiday in relaxed and solitary botanizing but, as it was, I was driven for miles and allowed to alight only occasionally; when I got back, my notebook was full. The most striking outcome was that within reasonable distance of that neat little farm, nothing came even close to the composition of the vegetation surrounding the crime scene in England.

I had accomplished all that I could except one thing. For elimination purposes, I needed to have some footwear from the family so that I could compare its botanical profile with that collected from the arrested gang members, the main suspect, and the crime scene itself. With the translator in tow, Dougie asked the family if he could buy a pair of shoes, sneakers, or boots normally worn by the family during their daily routine in and around the farm. A grubby pair of shoes was eventually proffered by the main suspect's sister and she was compensated with a sum that might have bought her a pair of Manolo Blahniks or Jimmy Choos. But I now had another set of critically important samples to process and analyze. Could I eliminate any of the suspected exhibits? Were any of the exhibits sufficiently similar to those of the comparator samples that they were implicated, meaning that even more scrutiny would be necessary?

The initial scans proved invaluable. A relatively high proportion of footwear bore no resemblance to the crime scene and could be disregarded. The state of the pollen and spores themselves and the background material on the preparation were both useful in eliminating samples. If all the grains are pristinely preserved in one soil, and badly corroded in another, it is unlikely that the source is common to both. The presence of fly-ash particles from combustion engines, or masses of fungal hyphae, or residual cellulose and lignin, can all help to characterize a sample. Pollen and spores are not the only important identifiers of place—the background "grot" on a slide can be very informative.

Many did not need full analysis, but some certainly did indicate temperate, mixed deciduous woodland with bracken fern. Each of these then had to be subjected to laborious counting.

To my amazement, the samples from the vehicle had obviously retained traces from a place very like the crime scene, and few, if any, like any of the places I had visited in Albania. I was worried, though—the picture was becoming bizarre. It was the footwear sold to us by the suspect's sister that was puzzling. How on earth could it yield results that were similar to those from the car and the crime scene? The pattern of trace markers on the shoes was not to be found in any of the Albanian soils with which she would have had daily contact. More work was needed. I asked for another pair of Albanian shoes to be brought over to the UK, and another pair was purchased from another member of the family. This pair had obviously picked up its pollen and spore load from the suspect's father's garden and there was no overall similarity to the Hertford-shire woodland.

When I reported this to detectives, after inquiries, they came back with an intriguing bit of information. The sister had said that her brother had given her the shoes that the police had originally bought—the final jigsaw piece fell into place. The brother seems to have worn the shoes to the grave site and they had retained the woodland profile, even such a long period after the offense. I sus-pected that after he had given them to his sister, she had only worn them around the house, or not worn them at all. If this were not the case, I would have expected to see more traces of the Al-banian farm on them. One can only surmise but, after leaving the woodland of the crime scene, the suspect probably got into his car, drove to his home in London and, soon after, left for Albania. I know from my own experience that when driving long distances over Europe, and eager to reach a destination, one's feet usually tread on urban streets, paving, and tarmac, then carpets or indoor flooring. None of these offer rich palyniferous surfaces and the original profile could be preserved in the nooks and crannies of-fered by virtually any kind of footwear.

I produced a matrix of results for the police and, although they were fragmentary in some respects, it could be seen clearly that footwear from several of the main suspect's associates had either been to the crime scene or somewhere very like it. The biological trace evidence picked up from the woodland had been carried back to the foot pedals and footwells of the very car that had transported the victim to his execution. Furthermore, the large number of items showing links with the woodland site could be explained; there were several illegal Albanians involved in the murder and, because the grave had been dug long before the execution was carried out, and then had to be re-excavated to provide a grave, more than one person had been there at least twice. So, there were several pairs of footwear and a car that could be linked to the suspect, his associates, and the crime scene. No one could claim that the pollen and spores had been transferred to the car and shoes in Albania because the profiles just did not match the soils there.

The suspect was already in prison in Tirana and, as nearly all the witnesses were unhappy about going back there again, the trial was held in the UK, but with four Albanian judges presiding. I was already standing to give evidence as the four judges entered the room. The demeanor of three, one of whom was a woman, was severe and serious, but one, in a shiny blue suit and rather flashy tie, was smiling and seemed full of charm. I later learned that he was the senior judge. All four were broad and squat, dressed in an exceedingly old-fashioned style, and nodded frequently as proceedings progressed. More and more, the unsmiling triumvirate became tough, no-nonsense apparatchiks.

Giving my evidence was slow and sheer tedium because everything had to be translated in detail, and very accurately. I was subjected to a slow-motion bombardment of questions. It was obvious that palynology, botany, and soils being used as evidence was quite outside any of the judges' experience. The verdict was "guilty," and a lengthy sentence was doled out in spite of the absence of a jury—but there was something wrong. It was only after I had escaped the courtroom that it dawned on me: I had been examined, but not cross-examined. There did not seem to be a defense lawyer. Later I was told that one of the male judges was acting on behalf of the defendant—but he had not challenged me with a single question. This was surreal. The evidence against the illegal Albanian immigrant was strong in palynological terms, but I came away feeling decidedly uncomfortable; the conviction seemed too easy. My skills have been honed by being battered and bruised in court, and having to engage in mental gymnastics to counteract the tactics of skilled lawyers. No judicial decision in the UK would ever have been achieved as easily as that.

I thought back to the little island in the stream gurgling through that quiet woodland. The murderer had planned his retribution, not doled it out in anger. The grave site had been chosen carefully so that it would be easy to find again, excavated many months in advance of the murder, and prepared specifically for that purpose. It must have seemed remote and safe to the killer, but he had not reckoned with the ubiquity of the British dog walker.

Even so, as I drove away from the court, my mind kept drifting back to the prison in Tirana and to all those women with baskets of food for their loved ones incarcerated inside. How would our illegal immigrant fare? I suspect not much better than the little cat.

CHAPTER 14

Endings

Reading books is never enough. The most important reservoir of information for an ecologist is the field. I have many exhilarating memories of lying in the heather, or long grass, watching and thrilling at the sound of skylarks rising almost vertically from the moorland, singing at great height before plunging dramatically back to Earth. The sight of oceans of purple moor-grass forming synchronous waves in response to the breeze moving over it, and the warmth of the heather bushes in full sun, with the buzz and hum of foraging insects all around me. There were also times when I sweated inside waterproof gear, horizontal rain penetrating every little gap, hair plastered to the scalp, sodden socks squelching in boots, and soggy, illegible notes. But how else could one see things in real life?

The key to interpretation is experience of real places and honing and crafting one's skills by walking, scrambling, climbing, and wading through bogs, ditches, fields, and woodlands. I spent years visiting archeological excavations from Hadrian's Wall to Pompeii and managed to recover valuable samples from pits, ditches, or anything that might provide information. They are messy places, but there were always keen young people wanting to learn a new skill and by teaching and supervising them I could get what I wanted without getting too dirty myself. I know that I have been the source of mirth on sites by arriving in a white sweater and it being just as pristine when I left.

By the time I found out about the others who had done some forensic palynology work, I had, in isolation, already invented my own subdiscipline of forensic ecology. I had worked successfully on several cases, established a code of best practice for various aspects of the work, and was well on the way to writing up my work for publication. Then I found out about others in the field; I was very pleased and eventually managed to contact them. It was a surprise to me that a colleague in the UK, Tony Brown, whom I had known for many years, had done some cases but was a busy university professor with other fish to fry. Looking back, it is sad that we did not get together and join forces. Another, Dallas Mildenhall, was operating in New Zealand and had been involved in forensic work for several years. Then, through him, I discovered a professor in Texas, Vaughn Bryant. What was so interesting is that none of us operated in the same way, or on the same kind of cases. Over the years, we have become close pen pals and we help each other as much as we can. I have visited Dallas in New Zealand and he has stayed with me several times in the UK. I have never met Vaughn but I feel he is a friend whom I know very well as we write to each other regularly.

I found that I was the only one who regularly attended crime scenes and mortuaries. I had just taken it for granted that if the police wanted to glean as much information as humanly possible from a crime scene, I had to be there quickly before anything could change or be contaminated. Being naturally quite forceful, I had always insisted on being "hands on" from beginning to end and, in this way, built up a large network of police colleagues and, indeed, forensic practitioners in very many areas of forensic science. The police rarely argued and they accepted my advice; in this way, they facilitated the development of the science, some to a considerable degree. My colleagues abroad never seemed to manage this kind of relationship with investigators, and I suspect this is partly due to the differences in the ways police services are organized and set up in those countries. Perhaps our characters and personalities had been an influence too. I seemed to be the only one with a strong enough stomach to face the most horrific of sights, smells, and the awfulness of corpses in every conceivable state of decay and mutilation. I always looked upon the deceased as a valuable source of evidence and developed techniques to maximize evidence retrieval from them. Importantly, I also had a sufficiently robust character to endure hostile police officers, attorneys and lawyers, as well as cope with the grim and worrisome character of the UK's Crown Courts.

How did I get to this point? When I look back, so much is a blur. I have never, ever, planned any aspect of my life and, although many cannot believe it, I am not a woman of ambition. It has all just happened to me; I have been reactive rather than proactive. I was already preadapted to become a forensic ecologist and had the necessary skills and knowledge from the laboratory, the field, and the literature. I had spent much of my academic life in university teaching and it is a truism that one learns a great deal from one's students. For the last six years that I was full-time at the University of London, I ran an MSc course on forensic archeological science. As well as the rigor of the theory, the course had a strong practical element and I held a class rota. As the turn of each pair of students came around, they could accompany me to the crime scene and/or the mortuary to act as my assistants. They had firsthand experience of real life and this certainly sorted out those who "could" from those who "could not." Teaching the MSc was challenging and fun for me, and I think the students felt that too. I look back now and it all seems such a long time ago. I am still giving lectures at universities in the UK and have traveled the world, running workshops and giving lectures in twenty-three countries. In the last couple years I have been to China twice, Peru, Colombia, and India.

It has been an incredibly full life from beginning to now, and it is still too full. I have little time to indulge myself in what I want to do—playing the piano more, sewing, craftwork, painting, cooking, and gardening—all rather solitary pastimes but all with a definite outcome. They are antidotes to the flurry of meetings that make up so much of my life. Every day I tell myself that I am going to spend the whole day on some indulgence, but usually end up at my computer screen, dealing with some problem or other. But I still have work to complete and publish. As one gets older, as well as funerals, one collects responsibilities—editing papers for journals, reviewing books and papers, checking other people's texts, and for me, being in local government. I was pleasantly surprised to get such a large majority in local elections as an independent councilor and am now a cabinet member in our local district council. I could never stand under the banner of one of the national political parties because I want to express my own views and reflect those of the electorate. It is difficult to juggle all the work, especially as I have the portfolio for "Environmental Health and Services." This seems to cover everything from trash cans, licensing, pollution of air, water, and soil to pigeon fouling—and everything in between. One certainly gets to know people.

I hated my grammar school but it seems to have endowed me with an education and skills that made me interesting enough to be the subject of broadcasts many times on radio and television. It even taught me to curtsy, something that came in useful when I was invited to lunch with the Queen and Prince Philip. That school also instilled in its pupils an overexaggerated sense of duty and responsibility. I regularly meet up with my old school chums and, after all these years, they seem just the same. We are all from an age of innocence and, thankfully, seem to have retained it. The contrast between my own life and those of the criminals and victims who have impinged on me over the years, is stark. I cannot help feeling sympathy for certain of them. Some have had a rotten start in life while others have not had the ability to grasp the endowment of riches gained from education. Some are victims of their families and circumstances, the people they fall in with, or unsympathetic authorities. Some just seem to be foul and evil.

Over the last ten years or so, government changes have transformed the forensic landscape in the UK. We used to have the government-run Forensic Science Service (FSS), which dealt with nearly all the police cases as a matter of course. Then a decision was made to allow competition and some ex–Forensic Science Service staff started up their own companies, first of all to provide good defense facilities for those faced with prosecution, but later expanding into prosecution work. These early entrepreneurs worked hard at absorbing any new discipline and skill to add to the base of standard scientific techniques used routinely. They commissioned people in the universities and anyone who had any unusual or rare skill that became feasible in forensic analysis.

In its wisdom, the government then decided that the official Forensic Science Service had to become competitive and would become a public company but with the government as the main shareholder. Oh dear, what a catastrophe. Essentially they were trying to commercialize a Civil Service Operation, but the FSS, with all the red tape typical of a government department, just could not respond quickly to change. The scientists dissipated and many set up their own companies. What is the situation now? We have two or three large commercial providers who aim to be one-stop shops to police forces that have been tied into contracts with them.

It is obvious that these providers cannot possibly afford to have the full range of skills in-house, so they commission anyone they think can serve them. On paper, the police can just check a box to say that they have a certain skill in place for an inquiry. But the quality of the skill? That is another matter. I certainly know that some with minimal experience have been employed on occasion, and I have, several times, been the cause of evidence being withdrawn once my evaluation and submission of the evidence has been delivered to the defense attorneys. Some of the reports I have seen on forensic palynology and soil analysis have been what can only be described as bizarre. The police have also decided to take on their own forensic work and some now employ ex-FSS employees in their own laboratories. One can imagine that there might just be a prosecution mentality there somewhere. I wonder if the in-house scientists are ever exposed to pressure to get the "right answer." Officers can be so convinced of guilt, and so keen to obtain a conviction, that one might fear the possibility of cognitive bias. One can only hope not. All forensic scientists need to be trained in providing unbiased reporting.

The American television program *CSI* became very popular. Of course much of it is fantastical, and the crime scene protocols, laboratory analysis, timelines, and results are often unrealistic and silly. But, the program certainly caught the public's imagination and universities were quick to realize that money was to be made by putting on, say, forensic chemistry rather than just chemistry. Forensic has become a very sexy word and more "forensic scientists" are being produced than would ever find a job. At the time of writing, several of my peers and I find the whole forensic scene in the UK somewhat depressing. There are hundreds of different university courses with the word "forensic" in the title. What about basic science—botany, zoology, chemistry, biochemistry, and mathematics? It is now virtually impossible to complete a botany degree in the UK, whereas the subject is seen as essential overseas and certainly in China, India, and even Spain. More and more we have to recruit for senior jobs in botanical science from overseas.

Everyone now learns about DNA but, even here, serious mistakes can be made. A few have arisen at the laboratory stage but this kind of error is much rarer than those of misinterpretation. There have been some serious miscarriages of justice in recent years because of inadequate interpretation of DNA results, and this is frightening. Even worse, the advancement of DNA technology has made it a victim of its own success. Techniques are now so highly sensitive that a DNA profile can be obtained from just a few cells. Imagine the problems when there are mixed profiles or contamination. Even if the laboratory technique is impeccable, the presence of a person's DNA at a crime scene does not necessarily mean that they had been present because of the ease with which DNA can be transferred. One of the great problems is understanding when there has been transfer from a primary source to a secondary one, and even to a tertiary one. When we meet our friends, or are compulsorily pressed up against people in the underground train, we are constantly exchanging DNA, so it is easy to see how some innocent person could be implicated in a crime. DNA results are also subjected to complex statistical techniques, and computer programs can be used to execute the analysis of the data, but they create their own problems. The person who has to interpret a DNA profile may not know what is going on within the analysis of the data. There is little doubt that DNA technology is highly advanced, and it exercises the mind of very many extremely clever people, but it is recognized that it does not always provide results that can be accepted with 100 percent confidence.

Forensic palynology is invariably needed when DNA or fingerprint evidence is of no help. Fingerprints, DNA, and fibers are three of the most important kinds of trace evidence in forensic science. Pollen and spores, and other microscopic particulates, provide another and, when produced and interpreted properly, the evidence they provide can be very powerful indeed. But I wonder whether my kind of work will survive and be part of forensic armory in the future? There are no broad-based botanists or mycologists now being trained in the UK's universities.

I frequently receive e-mails and letters from both students and experienced scientists from overseas asking if I will mentor them, or whether they could come and work with me so that they can learn the ropes. But when you point out that they would need palynological or botanical skills to PhD level and already have many years of work experience, the requests fade. I had been working as a researcher and lecturer for many years before I became so unexpectedly involved. I was already poised to be able to work out a strategy and execute it right through to court level. I was not a "learner," I simply amended what I already knew. My colleagues in Texas and New Zealand have also been scientists all their lives and came into the forensic arena late in their careers. They are well past retirement age and, although I find it hard to believe, so am I. This kind of work is not for the beginner.

I will never be able to look upon myself as being retired. To me, that is a bizarre concept. If there are things to do, do them. I have knowledge and skills that are useful, and it would be bad not to share them if needed. Sometimes I feel that I have done quite enough, and although I still take on cases, I never intend to give my life over to working with the police at the same intensity ever again. Life? I had no life. I did not even know my neighbors well and, in reality, was dedicated to working, much to the chagrin of my very few close friends.

One fascinating aspect of forensic ecology, or any kind of ecology, is that one never, ever stops learning. Every sample brings some surprise or other, and the rush of adrenaline that accompanies the trip from the slide tray to the microscope stage never fails to excite. It drives one on to investigate more. There is always more to look at, record, measure, interpret; the natural world seems infinite. In the first part of my career, I was seeking the past, and this meant digging into sediments and soils. Forensic work is mostly about the present and the evidence is mostly at the surface. The only thing that is predictable is the unpredictability of Mother Nature. One can draw up broad protocols and guides to best practice for carrying out a particular job, but one can never construct an infallible model to be applied from one scenario to another. Everything must be interrogated afresh.

A common fantasy among those of a certain age is having the opportunity to live one's life over again. Increasing years bring home very forcibly that you only have one go at it. This has become a cliché but, like so many clichés, it is so very true. I have never planned anything and my life just happened to me, but knowing what I know now, would I have done anything differently? Yes, I would. When I was a dreamy youngster, I thought I might be a ballerina, a concert pianist, or a research scientist. I suppose I achieved the lesser of the three, but looking back on my life, perhaps I should have gone into law and become a Queen's Counsel—a silk. I wouldn't want to be an ordinary lawyer as, irrespective of seniority, they are always referred to as being juniors. It is the QC who leads a case and has the responsibility of influencing a jury to give a verdict of innocent or guilty. In the UK, the US, and much of the Commonwealth, the legal system is adversarial. Both sides are in it to win and, in my experience, no holds barred. The prosecution puts the police case against the defendant and the defense attempts to act as a shield against any probing questions. You know you have done well, and that your evidence is strong, if the opposite side tries to denigrate you as a person. They have lost the plot at that point.

I have always loved court arguments where evidence is minutely scrutinized and challenged. The need to "think on one's feet"; the lawyer attacks, the witness parries, the lawyer then ripostes. Sometimes these games proceed for some time, depending on the confidence and robustness of the witness. Other times, a clever lawyer can destroy a witness with his opening statement.

Luckily that has not happened to me, but an experienced expert witness always expects the worst. The best lawyers are those who do their homework, or at least they have diligent juniors who do it for them. They are then exquisitely prepared for the sword to strike home. But, if you are truly expert, they can never do sufficient research to match your knowledge, and it never fails to amaze me that, in many cases, the lawyer has not been particularly well prepared at all.

Some seem to have a vague idea of what the case is about and present their "evidence in chief" based on what they can draw out of a prosecution witness. The best ones have been where, after submitting my reports, the attorney has requested meetings and spent time with me before going into court. The worst experience was when the QC had not read any of my reports at all before I made my oath, while the shortest one was in a murder case in the Old Bailey. I had to explain the complexities of my findings to the attorney in the public hall five minutes before going in to be cross-examined. I was perplexed and angry at this as I had done months of work; and the attorney would not be going to prison as a result of his lack of diligence—that would be the privilege of the defendant.

I have witnessed the outcomes of many different kinds of death—strangulation, poisoning, stabbing, suffocation, and mutilation as well as the outcomes of body disposals in various places and conditions. One thing that has always impressed me is that the body is an empty vessel; there is nothing left inside to make the body a person. There is no doubt in my mind that a someone becomes a something. We throw an empty bottle into the recycling bin without a thought other than protecting the planet from being polluted, and there have been events from prehistory through to today where human corpses have been treated in the same way. Our social norms demand that we engage in complex rituals when disposing of our dead, and these are observed irrespective of the closeness of relationship to the deceased.

I have lost a number of good friends in the last couple of years and, as I have stood in somber silence, mouthed hymns, and bowed my head in simulated prayer, I have often asked myself how much concern I had about that body in the coffin. Caring for memory of the person, our past relationship and times together is taken for granted, but, the body? Not much, but there are exceptions. Irrationally, I cared desperately about the fate of the bodies of my child, my grandmother, and every pet cat I have had. Why? I suppose I knew their bodies, their smells, and their feel intimately, and they were all precious. These feelings are irrational but I cannot deny them.

After death, the body will break down into the molecules which built it up from the food taken in. That person had converted the molecules of other organisms (meat and two veg) into those of his or her own, and these will now be released once more, to be taken up by others and perpetuate the cycle of life. A body left on the surface will break down much more quickly than a buried one and a cremated individual will be reduced to mineral ash within minutes. If the ash is spread about in a woodland, that person will be truly reincarnated. Elements in the ash will be taken up by bacteria, fungi, invertebrates, and plant roots. One individual can spread throughout a woodland and become many. How wonderful to be reincarnated as a bluebell, an oak tree, and a lovely beetle all at the same time. It will certainly happen whether you like the idea or not.

I find this concept very appealing and I know that my husband's molecules and mine will mingle. Our ashes will be spread in the same place so we might even both end up in the same tree or bluebell. How marvelous! When the tree or bluebell dies and their corpses decompose, our molecules may be released again and taken up by yet other living things. The elements that make up our bodies will exist as long as the Earth revolves around the Sun.

The sad thing for me is that I will not be able to know about any of it. I will cease to be. I will not have a physical monument, and I am not so vain that I think anyone will remember me after my nearest and dearest have gone. There will be no musings in a churchyard, or even a municipal cemetery. I am not a lover of poetry, although I always found Gray's *Elegy* quite emotional, but my gravestone will not exist to be able to move anyone to write anything. I suppose my monument will be my work and my publications. My words will live on, and rather than in a testament to sentimentality, which graveyards certainly are, evidence of my existence will probably be found in some dusty old library somewhere.

I am often asked if my experiences with death, rape, and other crimes have affected me. The two deaths that have affected me most in my life were those of my daughter and my grandmother. I still miss my grandmother for her wisdom and comforting presence. My daughter is, and always will be a constant ache, kept deep inside, but visited every day, even after all these years. There is little doubt that their deaths have made me realize that every corpse I have encountered also probably had someone who felt the same about them, and this keeps me respectful and caring, certainly for innocent victims anyway. Although the body on the table means little to me, it does to someone else, and this must be kept in mind at all times. One must remain objective or else it would be difficult to do useful work but one must not forget that a corpse was a person.

What were the lessons I learned during this life so impinged upon by crime? I certainly learned a few swearwords and I discovered how to keep a vacant expression when my insides were in turmoil. I suppose I have learned to be utterly pragmatic when faced with a problem, and to apply Occam's razor when those difficulties become intractable. I have had the reputation for being somewhat hard and unsympathetic, although those who really know me understand that I am soft inside. I do not like hurting anyone or anything and certainly will not kill anything unnecessarily. I am very sharp with those who are lazy, dishonest, selfish, and manipulative, and I try to be utterly honest myself. I suppose I genuinely want to be a good person, but whether I achieve my aspiration is for others to say. I hope that people will remember me for being meticulous, hardworking, helpful and, most of all, kind—not much of an epitaph, but respectable enough.

I wonder what my death will be like? I am certainly not going to have the usual mournful funeral but will try to put together a PowerPoint presentation of my goodbyes, as soon as I get a hint that I am on my way out. I only hope that my gray cells will still be working, that I will die with minimal pain and discomfort, in my own home, in my own bed, in the arms of my darling husband who, ever since I met him, has been my rock and my joy in life.

ACKNOWLEDGMENTS

There are many people to whom I owe a debt of gratitude in my quest for excellence in forensic ecology. The list is long but I must first acknowledge my dear husband, Professor David L. Hawksworth, CBE, who has encouraged and helped me in every way. Then, how could I not give credit to Dr. Judy Webb, who worked with me for so many years, and whose brilliance in pollen identification has enhanced the development of forensic palynology. Thanks too to Professor Kevin Edwards, my longtime friend, colleague, and sternest critic, who has helped me to maintain high standards, and Peter Murphy, the dearest of friends and colleagues who, in spite of his misgivings about being involved, made so much of my work tolerable. I am indebted to the Institute of Archaeology, University College London, for giving me such good facilities, including technical help by the wonderful and knowledgeable Sandra Bond; and to my colleagues there, particularly Dr. Richard Macphail and the late Professor Gordon Hillman, for their intellectual support and lots of fun. I cannot forget my magnificent experiences at King's College London, both as a student and lecturer. Many doors were opened for me by the inspired teaching and kindnesses I received from the staff, particularly from Dr. Peter Moore, Professor Bill Bradbeer, the late Dr. Francis Rose, and the late

Professor Arthur Bell, who continued his help when he became Director of the Royal Botanic Gardens, Kew. Thanks too to Professor Frank Chambers and Dr. John Daniell at the University of Gloucestershire, and Professor Tony Brown at the University of Southampton, who have all facilitated my work. Then there are so very many students, some of whom were utterly brilliant, but all of whom taught me a lot. Not least, I owe so much to all the clever and astute policemen with whom I have worked over the years, especially Detective Chief Superintendent Paul Dockley, who gave me my first job, Detective Sergeants Bill Bryden, MBE, and Doug Bain, who showed such confidence in my work, as well as Detective Superintendent Ray Higgins, the kindest of men.